LONDON MATHEMATICAL SOCIETY LECTURE NOTE SERIES

1. General cohomology theory and K-theory, P.HILTO
4. Algebraic topology: a student's guide, J.F.ADA
5. Commutative algebra, J.T.KNIGHT
8. Integration and harmonic analysis on compact groups, R.E.EDWARDS
9. Elliptic functions and elliptic curves, P.DU VAL
10. Numerical ranges II, F.F.BONSALL & J.DUNCAN
11. New developments in topology, G.SEGAL (ed.)
12. Symposium on complex analysis, Canterbury, 1973, J.CLUNIE & W.K.HAYMAN (eds.)
13. Combinatorics: Proceedings of the British combinatorial conference 1973, T.P.McDONOUGH & V.C.MAVRON (eds.)
14. Analytic theory of abelian varieties, H.P.F.SWINNERTON-DYER
15. An introduction to topological groups, P.J.HIGGINS
16. Topics in finite groups, T.M.GAGEN
17. Differentiable germs and catastrophes, Th.BROCKER & L.LANDER
18. A geometric approach to homology theory, S.BUONCRISTIANO, C.P.ROURKE & B.J.SANDERSON
20. Sheaf theory, B.R.TENNISON
21. Automatic continuity of linear operators, A.M.SINCLAIR
23. Parallelisms of complete designs, P.J.CAMERON
24. The topology of Stiefel manifolds, I.M.JAMES
25. Lie groups and compact groups, J.F.PRICE
26. Transformation groups: Proceedings of the conference in the University of Newcastle upon Tyne, August 1976, C.KOSNIOWSKI
27. Skew field constructions, P.M.COHN
28. Brownian motion, Hardy spaces and bounded mean oscillation, K.E.PETERSEN
29. Pontryagin duality and the structure of locally compact abelian groups, S.A.MORRIS
30. Interaction models, N.L.BIGGS
31. Continuous crossed products and type III von Neumann algebras, A.VAN DAELE
32. Uniform algebras and Jensen measures, T.W.GAMELIN
33. Permutation groups and combinatorial structures, N.L.BIGGS & A.T.WHITE
34. Representation theory of Lie groups, M.F.ATIYAH et al.
35. Trace ideals and their applications, B.SIMON
36. Homological group theory, C.T.C.WALL (ed.)
37. Partially ordered rings and semi-algebraic geometry, G.W.BRUMFIEL
38. Surveys in combinatorics, B.BOLLOBAS (ed.)
39. Affine sets and affine groups, D.G.NORTHCOTT
40. Introduction to H_p spaces, P.J.KOOSIS
41. Theory and applications of Hopf bifurcation, B.D.HASSARD, N.D.KAZARINOFF & Y-H.WAN
42. Topics in the theory of group presentations, D.L.JOHNSON
43. Graphs, codes and designs, P.J.CAMERON & J.H.VAN LINT
44. Z/2-homotopy theory, M.C.CRABB
45. Recursion theory: its generalisations and applications, F.R.DRAKE & S.S.WAINER (eds.)
46. p-adic analysis: a short course on recent work, N.KOBLITZ
47. Coding the Universe, A. BELLER, R. JENSEN & P. WELCH
48. Low-dimensional topology, R. BROWN & T.L. THICKSTUN (eds.)
49. Finite geometries and designs, P. CAMERON, J.W.P. HIRSCHFELD & D.R. HUGHES (eds.)
50. Commutator Calculus and groups of homotopy classes, H.J. BAUES
51. Synthetic differential geometry, A. KOCK
52. Combinatorics, H.N.V. TEMPERLEY (ed.)
53. Singularity theory, V.I. ARNOLD
54. Markov processes and related problems of analysis, E.B. DYNKIN
55. Ordered permutation groups, A.M.W. GLASS
56. Journees arithmetiques 1980, J.V. ARMITAGE (ed.)
57. Techniques of geometric topology, R.A. FENN
58. Singularities of differentiable functions, J. MARTINET
59. Applicable differential geometry, F.A.E. PIRANI and M. CRAMPIN
60. Integrable systems, S.P. NOVIKOV et al.

61. The core model, A. DODD
62. Economics for mathematicians, J.W.S. CASSELS
63. Continuous semigroups in Banach algebras, A.M. SINCLAIR
64. Basic concepts of enriched category theory, G.M. KELLY
65. Several complex variables and complex manifolds I, M.J. FIELD
66. Several complex variables and complex manifolds II, M.J. FIELD
67. Classification problems in ergodic theory, W. PARRY & S. TUNCEL

London Mathematical Society Lecture Note Series. 67

Classification Problems in Ergodic Theory

WILLIAM PARRY and SELIM TUNCEL
Mathematical Institute
University of Warwick

CAMBRIDGE UNIVERSITY PRESS

CAMBRIDGE

LONDON NEW YORK NEW ROCHELLE

MELBOURNE SYDNEY

CAMBRIDGE UNIVERSITY PRESS
Cambridge, New York, Melbourne, Madrid, Cape Town, Singapore, São Paulo

Cambridge University Press
The Edinburgh Building, Cambridge CB2 8RU, UK

Published in the United States of America by Cambridge University Press, New York

www.cambridge.org
Information on this title: www.cambridge.org/9780521287944

First published 1982
Re-issued in this digitally printed version 2008

A catalogue record for this publication is available from the British Library

Library of Congress Catalogue Card Number: 82–4350

ISBN 978-0-521-28794-4 paperback

CONTENTS

	Page
Preface	
Chapter I: Introduction	1
1. Motivation	1
2. Basic Definitions and Conventions	2
3. Processes	3
4. Markov Chains	4
5. Reduced Processes and Topological Markov Chains	5
6. Information and Entropy	7
7. Types of Classification	10
Chapter II: The Information Cocycle	12
1. Regular Isomorphisms	12
2. Unitary Operators and Cocycles	14
3. Information Variance	18
4. The Variational Principle for Topological Markov Chains	21
5. A Group Invariant	26
6. Quasi-regular Isomorphisms and Bounded Codes	31
7. Central Limiting Distributions as Invariants	36
Chapter III: Finitary Isomorphisms	38
1. The Marker Method	38
2. Finite Expected Code-lengths	50
Chapter IV: Block-codes	54
1. Continuity and Block-codes	54
2. Bounded-to-one Codes	56
3. Suspensions and Winding Numbers	59
4. Computation of the First Cohomology Group	62

Chapter V: Classifications of Topological Markov Chains 64

1. Finite Equivalence 64
2. Almost Topological Conjugacy and the Road Problem 69
3. Topological Conjugacy of Topological Markov Chains 72
4. Invariants and Reversibility 79
5. Flow Equivalence 83

Appendix: Shannon's Work on Maximal Measures 90

References 96

Index 100

PREFACE

These notes grew out of M. Sc. lecture courses given by the first named author at the Mathematics Institute (University of Warwick) in 1976 and 1977. To be more precise, the material presented here is concerned only with those parts of the courses which are not to be found in well known texts, together with additional material, organised by the second named author, which was not presented in the above courses.

The M. Sc. audience was expected to be familiar with measure and integration theory and it was assumed that students had had at least some contact with elementary functional analysis including Hilbert space theory. The foundations of the course, which were discussed in an informal tutorial fashion, consisted roughly of the following topics: measure-preserving transformations, recurrence, Birkhoff's and von Neumann's ergodic theorems, conditional expectation, increasing and decreasing sequences of σ-algebras and the associated Martingale theorems, information, entropy and the Shannon-McMillan-Breiman theorem. Students were expected to read these topics as an integral part of the course and were advised to refer to the relevant sections of [H] , [R. 2], [W. 1] (to which we would now add [P. 1]). Readers of these notes who are unfamiliar with these foundations are similarly advised.

We offer our thanks to M. Keane and M. Smorodinsky for a number of consultations concerning Chapter III. We would also like to record our gratitude to Klaus Schmidt and Peter Walters for helpful critical comments concerning an earlier draft of these notes.

CHAPTER I: INTRODUCTION

1. MOTIVATION

The problems discussed in these notes are motivated by the classical isomorphism problem of ergodic theory, which has received much attention since the introduction (by Kolmogorov) of entropy into the subject, and especially since Ornstein's celebrated solution for Bernoulli automorphisms. According to Kolmogorov, two Bernoulli automorphisms S, T (shifts on product spaces $(X, m) = \prod_{-\infty}^{\infty} (X_0, m_0)$ and $(Y, p) = \prod_{-\infty}^{\infty} (Y_0, p_0)$, respectively, where (X_0, m_0) and (Y_0, p_0) are finite probability spaces) are isomorphic only if their entropies $(-\sum_{i \in X_0} m_0(i) \log m_0(i)$ and $-\sum_{j \in Y_0} p_0(j) \log p_0(j))$ coincide. Ornstein's result ([O.1], [O.2]) complements this statement: if two Bernoulli automorphisms have the same entropy then they are isomorphic. This means that there is an essentially invertible measure-preserving transformation ϕ from almost all of X onto almost all of Y satisfying $\phi S = T\phi$ a.e. Writing $\phi(x) = (\phi_n(x))$ in terms of component functions ϕ_n, we see that $\phi_n(x) = \phi_0(S^n x)$ and ϕ is determined by $\phi_0 : X \rightarrow Y_0$. The present, $\phi_0(x)$, may depend on the entire past and future of x; this is not satisfactory from the point of view of communication and coding.

We have indicated that isomorphisms, which preserve only the most basic structures of measure theory, may be too weak for applications. It is therefore appropriate to ask for measure-preserving transformations which respect structures other than just measure e.g. a sub-σ-algebra which represents the 'past' or a partition (finite or countable) which represents the 'state space' of a stationary stochastic process. Also, finite state processes have natural topologies so that it is natural to consider continuous maps and homeomorphisms between them. These considerations lead us to the concepts of regular isomorphism, quasi-regular isomorphism, finitary code, block code, finite equivalence, almost topological conjugacy and topological conjugacy. A more detailed preview of the ideas involved in these will be given in the final section of this chapter.

In many of the above classifications an information function can be used as an invariant. Indeed, the question of the extent to which information, as opposed to entropy, can be used in classification problems is a principal motivation for many of the results presented in these notes.

2. BASIC DEFINITIONS AND CONVENTIONS

As is usual in measure theory, all measure theoretic objects which are equal almost everywhere are identified throughout these notes. Thus the qualification 'almost everywhere' is often omitted. We always require sets and functions to be measurable, even when this is not explicitly specified. $(X, \mathcal{B}, m)$ or a similar triple always denotes a probability space. If $\mathcal{A} \subset \mathcal{B}$ is a sub-σ-algebra and $\mathcal{A}_n$ is a sequence of sub-σ-algebras such that $\bigcup_n \mathcal{A}_n$ generates $\mathcal{A}$ then we write $\mathcal{A}_n \uparrow \mathcal{A}$.

For probability spaces $(X_i, \mathcal{B}_i, m_i)$ $(i = 1, 2)$, a measure-preserving transformation ϕ from almost all of X_1 onto almost all of X_2 is called a homomorphism and in keeping with the above convention we write $\phi : X_1 \to X_2$. Measure-preserving means $\phi^{-1}\mathcal{B}_2 \subset \mathcal{B}_1$ and $m_1(\phi^{-1}B) = m_2(B)$ for all $B \in \mathcal{B}_2$. If $\phi : X_1 \to X_2$ is a homomorphism, $(X_2, \mathcal{B}_2, m_2)$ is called a homomorphic image or factor of $(X_1, \mathcal{B}_1, m_1)$ and $(X_1, \mathcal{B}_1, m_1)$ an extension of $(X_2, \mathcal{B}_2, m_2)$. ϕ is called an isomorphism if there is a homomorphism $\psi : X_2 \to X_1$ such that $\psi\phi = id_{X_1}$ and $\phi\psi = id_{X_2}$ (a.e.). A homomorphism (isomorphism) from a probability space to itself is called an endomorphism (automorphism). An endomorphism T of $(X, \mathcal{B}, m)$ is ergodic if $T^{-1}B = B$, $B \in \mathcal{B}$ implies $mB = 0$ or 1.

A probability space which is isomorphic to a subinterval of $[0, 1]$ (with Lebesgue measure) together with a countable number of atoms is called a Lebesgue space. All of our spaces will be Lebesgue. This is not a very restrictive condition; for instance a complete separable metric space together with a complete Borel probability always defines a Lebesgue space. The main reasons for imposing this condition are that Lebesgue spaces are separable and that the following holds (see [R. 1]):

Let $(X_1, \mathcal{B}_1, m_1)$ and $(X_2, \mathcal{B}_2, m_2)$ be Lebesgue spaces. Every σ-algebraic homomorphism $\Phi : \mathcal{B}_2 \to \mathcal{B}_1$ $(\Phi(B^C) = (\Phi B)^C$, $\Phi(\bigcup_n B_n) = \bigcup_n \Phi(B_n)$, $m_1(\Phi B) = m_2 B$ for all $B, B_n \in \mathcal{B}_2$ and $\Phi(X_2) = X_1)$ is realised, essentially uniquely, by a homomorphism $\phi : X_1 \to X_2$ in the sense that $\Phi B = \phi^{-1} B$ for all $B \in \mathcal{B}_2$.

For $i = 1, 2$ let T_i be an endomorphism of $(X_i, \mathcal{B}_i, m_i)$. A homomorphism $\phi : X_1 \to X_2$ satisfying $\phi T_1 = T_2 \phi$ is called a homomorphism of T_1 to T_2; T_2 is then a factor of T_1 and T_1 an extension of T_2. (We write $T_1 \xrightarrow{\phi} T_2$.) If such a homomorphism is an isomorphism, then T_1 and T_2 are said to be isomorphic or conjugate.

Let T be an endomorphism of $(X, \mathcal{B}, m)$. There is a 1-1 correspondence between T-invariant sub-σ-algebras $\mathcal{A}$ $(T^{-1}\mathcal{A} \subset \mathcal{A} \subset \mathcal{B})$ and factors of T: If $T' : (X', \mathcal{B}', m') \to (X', \mathcal{B}', m')$ is a factor of T by ϕ then $T^{-1}(\phi^{-1}\mathcal{B}') \subset \phi^{-1}\mathcal{B}' \subset \mathcal{B}$ and, on the other hand, if $T^{-1}\mathcal{A} \subset \mathcal{A} \subset \mathcal{B}$ then there exists a factor (by some ϕ) $T' : (X', \mathcal{B}', m') \to (X', \mathcal{B}', m')$, unique up to isomorphism and called the factor with respect to $\mathcal{A}$, such that $\mathcal{A} = \phi^{-1}\mathcal{B}'$. Moreover there is an automorphism $\hat{T}$ of a (Lebesgue) space $(\hat{X}, \hat{\mathcal{B}}, \hat{m})$, unique up to isomorphism and called the natural extension of T, such that $\hat{T}$ is an extension of T (by ϕ, say) and $\hat{T}^n \phi^{-1}\mathcal{B} \uparrow \hat{\mathcal{B}}$ (see [R. 3]).

If T is an automorphism of $(X, \mathcal{B}, m)$, an invariant sub-σ-algebra $\mathcal{A}$ with $T^n\mathcal{A} \uparrow \mathcal{B}$ is called exhaustive. An automorphism of a Lebesgue space together with a preferred exhaustive sub-σ-algebra is called a process.

3. PROCESSES

Let X_0 be a finite or countable set, and let $X = \prod_{n=-\infty}^{\infty} X_n$ where $X_n = X_0$ for all $n \in \mathbb{Z}$. The shift T is defined $(Tx)_n = x_{n+1}$ for $x = (x_n) \in X$. A cylinder is a set of the form

$$[i_0, i_1, \ldots, i_l]^k = \{x = (x_n) \in X : x_k = i_0, x_{k+1} = i_1, \ldots, x_{k+l} = i_l\}$$

where $k, l \in \mathbb{Z}$ and $l \geq 0$. If $k = 0$, we sometimes omit this superscript. Let $\mathcal{B}$ be the σ-algebra generated by all cylinder sets. Clearly $T^{-1}\mathcal{B} = \mathcal{B}$. If m is a T-invariant probability on $(X, \mathcal{B})$ then $(X, \mathcal{B}, m, T)$, or simply T, is called a countable or finite state process according to the cardinality of X_0, the state space.

Consider an automorphism T of a Lebesgue space $(X, \mathcal{B}, m)$. A finite or countable (measurable) partition α is called a generator if $\bigcup_{n=-\infty}^{\infty} T^n \alpha$ generates $\mathcal{B}$. If $\mathcal{B}$ is even generated by $\bigcup_{n=0}^{\infty} T^{-n}\alpha$, α is called a strong generator. The state space X_0 of a countable or finite state process defines a generator called the state partition: $\{[i] : i \in X_0\}$. In fact there is no significant difference between automorphisms with specified generators and (countable or finite) state processes, as the exercise below shows. This rests on the fact that we are restricting attention to Lebesgue spaces, since we need the following: If $(\mathcal{P}_n)$ is a sequence of partitions of a Lebesgue space $(X, \mathcal{B}, m)$ such that $\bigcup_n \mathcal{P}_n$ generates $\mathcal{B}$, then there is a null set N such that for all $x, y \in X - N$ there exist n and $B \in \mathcal{P}_n$ with $x \in B$, $y \in X - B$ (see [R. 1]).

Exercise. Let T be an automorphism of a Lebesgue space $(X, \mathcal{B}, m)$ and let $\alpha = \{A_1, A_2, \dots\}$ be a finite or countable generator. Let $X_0 = \{1, 2, \dots\}$ have the same cardinality as α and define $X' = \prod_{-\infty}^{\infty} X_0$. Define a probability μ on X' by

$$\mu[i_0, i_1, \dots, i_l]^k = m(A_{i_0} \cap T^{-1} A_{i_1} \cap \dots \cap T^{-l} A_{i_l}).$$

Note that the shift S preserves μ, and prove that S and T are isomorphic.

Given a countable or finite state process $(X, \mathcal{B}, m, T)$, let $\mathcal{A}$ be the 'past' sub-σ-algebra generated by the cylinders $[i_0, \dots, i_l]^k$ with $k \geq 0$. Then $\mathcal{A}$ is T-invariant and exhaustive. $\mathcal{A}$ is called the standard past, and $(X, \mathcal{B}, m, T)$ is understood to have $\mathcal{A}$ as its preferred exhaustive sub-σ-algebra.

4. MARKOV CHAINS

In this section we list some definitions and facts concerning non-negative matrices and Markov chains. The basic references are [S'] and [F].

Let A be a non-negative $k \times k$ matrix. A is irreducible if for each pair (i, j), $1 \leq i, j \leq k$, we can find $n \geq 1$ such that the product A^n has $A^n(i, j) > 0$. The period of (a state) i $(1 \leq i \leq k)$ is the highest common factor of $n \geq 1$ with $A^n(i, i) > 0$. If A is irreducible, this is independent of the state chosen and is called the period of A. A is called aperiodic if (it is irreducible and) it has period 1; this is equivalent to requiring $A^n > 0$ for some $n \geq 1$.

A non-negative matrix is stochastic if all its row sums equal 1. A probability vector is a strictly positive vector with sum 1. Given an irreducible stochastic $k \times k$ matrix P, there is a unique probability vector p with $pP = p$. Hence we obtain a finite state process $(X, \mathcal{B}, m, T)$ by taking $X = \prod_{-\infty}^{\infty} \{1, 2, \dots, k\}$ and defining

$$m[i_0, i_1, \ldots, i_l]^n = p(i_0) P(i_0, i_1) P(i_1, i_2) \ldots P(i_{l-1}, i_l).$$

$(X, \mathcal{B}, m, T)$, together with its state partition, is called the Markov chain defined by P. The (shift) automorphism T and the measure m are also called Markov. If P has identical rows and $pP = p$ as above, the unique probability vector p is easily seen to be the one giving the rows of P so that m is the product measure obtained from the measure on $\{1, \ldots, k\}$ which assigns $p(i)$ to i. In this case the process, shift and measure are called Bernoulli.

If P is an irreducible stochastic $k \times k$ matrix of period t and p is the unique probability vector with $pP = p$ then for each pair (i, j), $1 \leq i, j \leq k$, there exists $0 \leq t_0 \leq t-1$ such that $\lim_{n \to \infty} P^{t_0 + nt}(i, j) = tp(j)$ and $P^m(i, j) = 0$ if m is not of the form $t_0 + nt$. This shows that all Markov automorphisms are ergodic. The convergence of $P^{t_0 + nt}(i, j)$ is exponentially fast.

We will need:

Perron-Frobenius Theorem [S']. Let A be a non-negative irreducible matrix with period t. Then

(i) there is a positive eigenvalue β with a corresponding strictly positive eigenvector;

(ii) β is a simple eigenvalue (i.e. it is a simple root of the characteristic equation of A);

(iii) $\beta\omega^i$, $i = 0, 1, \ldots, t-1$ are eigenvalues where ω is a primitive t^{th} root of 1, and for all other eigenvalues α, $|\alpha| < \beta$;

(iv) if $r = (r_1, \ldots, r_k)$ is a strictly positive vector then

$$\min_{1 \leq i \leq k} \left\{\frac{(Ar)_i}{r_i}\right\} \leq \beta \leq \max_{1 \leq i \leq k} \left\{\frac{(Ar)_i}{r_i}\right\}$$

with equality on either side implying equality throughout. In particular $Ar = \alpha r$, r strictly positive, implies that $\alpha = \beta$.

5. REDUCED PROCESSES AND TOPOLOGICAL MARKOV CHAINS

We shall consider all finite state shift spaces, whenever necessary, as topologized in the following natural manner: If $X = \prod_{-\infty}^{\infty} X_0$ where X_0 is a finite set, we give X_0 the discrete topology and X the product topology. Then X is compact and metrizable and the shift is a homeomorphism. The closed-open subsets of X are the cylinder sets and finite unions of these. Cylinders form a base for the topology of X and so, generate the Borel σ-algebra. Moreover, X is zero-dimensional (it has a base consisting of open-closed sets).

Suppose $(X, \mathcal{B}, m, T)$ is a finite state process. Then $\mathcal{B}$ is the Borel σ-algebra. Let X' be the support of m i.e. $X' = X - U$ where U is the largest open null subset of X. X' is T-invariant. We denote by $\mathcal{B}'$, m', T' the restrictions of $\mathcal{B}$, m, T to X'. X' is compact and zero-dimensional and T' is a homeomorphism. $(X', \mathcal{B}', m', T')$ is the <u>reduced process.</u> Measure theoretically there is no distinction between a finite state process and its reduced process and, since supporting measures are more convenient for topological considerations, we shall always assume that <u>finite state processes are reduced.</u>

Suppose A is a 0-1 irreducible $k \times k$ matrix. A defines a closed, shift-invariant subset X of $\prod_{-\infty}^{\infty}\{1, 2, \dots, k\}$:

$$X = \{x = (x_n) : A(x_n, x_{n+1}) = 1 \text{ for all } n\}.$$

X, together with (the restriction of) the shift, is called the <u>topological Markov chain</u> or <u>subshift of finite type</u> defined by A. If P is a $k \times k$ (irreducible) stochastic matrix compatible with A (i.e. $P(i, j) = 0$ iff $A(i, j) = 0$), the Markov measure defined by P has X as its support. Our assumption that finite state processes are reduced means that we regard Markov measures as being defined on their supporting topological Markov chains. The period of a topological Markov chain is defined according to the period of its defining matrix A.

The 0 - 1 matrix A may be viewed as a matrix of transitions: we have k vertices and a transition from i to j is allowed iff $A(i, j) = 1$. The topological Markov chain given by A is the space of all doubly infinite sequences of (allowable) transitions. From this point of view, there is no reason to restrict ourselves to 0-1 matrices; given an irreducible non-negative integral $k \times k$ matrix A' we again have k vertices and $A'(i, j)$ specifies the number of paths from i to j. The topological Markov chain now consists of all doubly infinite sequences of directed paths, with the shift transformation. This, however, is no more general than the 0-1 case since we may index with the directed paths a 0-1 matrix (transition from path a to path b is allowed iff b starts at the terminal vertex of a) which gives an equivalent (topologically conjugate) topological Markov chain in the sense that there exists a homeomorphism between the two spaces which conjugates the shifts. We shall work with 0-1 matrices, resorting to general non-negative matrices only to save space in examples.

[A. M.] contains a beautiful exposition of topological Markov chains.

6. INFORMATION AND ENTROPY

In this section we give a brief review of the basics of information and entropy theory. Details may be found in [B], [R. 2] and [W. 1], but [P. 1] is perhaps the best reference for our purposes.

If α is a countable partition of the (Lebesgue) space $(X, \mathcal{B}, m)$ and $\mathcal{C} \subset \mathcal{B}$ is a sub-σ-algebra, the <u>conditional information of</u> α <u>given</u> $\mathcal{C}$ is

$$I(\alpha|\mathcal{C}) = -\sum_{A\in\alpha} \chi_A \log m(A|\mathcal{C}) .$$

All logarithms are to the base e. $H(\alpha|\mathcal{C}) = \int I(\alpha|\mathcal{C})\, dm$ is the <u>conditional entropy of</u> α <u>given</u> $\mathcal{C}$. If $\mathcal{C}$ is the trivial σ-algebra consisting of sets of measure 0 and 1, we have, respectively, the <u>information</u> and <u>entropy</u> of α, $I(\alpha)$ and $H(\alpha)$. Note that $I(\alpha|\mathcal{C}) \geq 0$ and that $H(\alpha|\mathcal{C}) = 0$ (or $I(\alpha|\mathcal{C}) = 0$) iff α consists of sets in $\mathcal{C}$.

For a countable partition α, we shall use the same symbol to denote the σ-algebra generated by the partition, the distinction being clear from the context. For partitions α, β we put $\alpha \vee \beta = \{A \cap B : A \in \alpha, B \in \beta\}$. We use similar notation for the refinement of any number of partitions. When $\mathcal{C}_1, \mathcal{C}_2, \ldots$ is a sequence of σ-algebras, $\bigvee_{n=1}^{\infty} \mathcal{C}_n$ denotes the σ-algebra generated by their union.

The <u>basic identities for information and entropy</u> are:

$$I(\alpha \vee \beta|\gamma) = I(\alpha|\beta \vee \gamma) + I(\beta|\gamma)$$

$$H(\alpha \vee \beta|\gamma) = H(\alpha|\beta \vee \gamma) + H(\beta|\gamma)$$

for countable partitions α, β, γ. These are easily verified. If $\mathcal{C} \subset \mathcal{B}$ is a σ-algebra we can find (finite) partitions γ_n such that $\gamma_n \uparrow \mathcal{C}$, since we are in a Lebesgue (therefore separable) space. Now using γ_n in place of γ in the basic identities, taking limits, and using the increasing Martingale theorem we obtain,

$$I(\alpha \vee \beta|\mathcal{C}) = I(\alpha|\beta \vee \mathcal{C}) + I(\beta|\mathcal{C}),$$

$$H(\alpha \vee \beta|\mathcal{C}) = H(\alpha|\beta \vee \mathcal{C}) + H(\beta|\mathcal{C})$$

for countable partitions α, β and a sub-σ-algebra $\mathcal{C}$. From these identities we see that $I(\alpha|\mathcal{C}) \geq I(\beta|\mathcal{C})$, $H(\alpha|\mathcal{C}) \geq H(\beta|\mathcal{C})$ when $\alpha \geq \beta$ (i.e. when $\alpha \supset \beta$ as σ-algebras generated by the partitions). That $H(\alpha|\mathcal{A}) \leq H(\alpha|\mathcal{C})$ when $\mathcal{A} \supset \mathcal{C}$ may be proved with the aid of Jensen's inequality, but the corresponding inequality for information is not generally true.

Suppose $\mathcal{A}$, $\mathcal{C}$ are sub-σ-algebras and let $\alpha_n \uparrow \mathcal{A}$, $\beta_n \uparrow \mathcal{A}$ where α_n, β_n are finite partitions (for $n = 1, 2, \ldots$). Then

$$I(\beta_m|\mathcal{C}) \leq I(\alpha_n \vee \beta_m|\mathcal{C}) = I(\alpha_n|\mathcal{C}) + I(\beta_m|\alpha_n \vee \mathcal{C})$$

and, letting $n \to \infty$,

$$\lim_{n\to\infty} I(\alpha_n|\mathcal{C}) + I(\beta_m|\mathcal{A} \vee \mathcal{C}) \geq I(\beta_m|\mathcal{C})$$

so that, since $\beta_m \subset \mathcal{A}$,

$$\lim_{n\to\infty} I(\alpha_n|\mathcal{C}) \geq I(\beta_m|\mathcal{C}).$$

Letting $m \to \infty$,

$$\lim_{n\to\infty} I(\alpha_n|\mathcal{C}) \geq \lim_{m\to\infty} I(\beta_m|\mathcal{C}).$$

Clearly the reverse inequality is also true, so that the definition

$$I(\mathcal{A}|\mathcal{C}) = \lim_{n\to\infty} I(\alpha_n|\mathcal{C}) \quad \text{for } \alpha_n \uparrow \mathcal{A}$$

is unambiguous. Note that $\lim_{n\to\infty} I(\alpha_n|\mathcal{C})$ exists in $\mathbb{R} \cup \{\infty\}$ as $I(\alpha_n|\mathcal{C})$ are increasing (for increasing α_n). Again we define $H(\mathcal{A}|\mathcal{C}) = \int I(\mathcal{A}|\mathcal{C})\,dm$. It should be clear that the following sharpened versions of the basic identities are valid:

$$I(\mathcal{A}_1 \vee \mathcal{A}_2|\mathcal{C}) = I(\mathcal{A}_2|\mathcal{A}_1 \vee \mathcal{C}) + I(\mathcal{A}_1|\mathcal{C}),$$

$$H(\mathcal{A}_1 \vee \mathcal{A}_2|\mathcal{C}) = H(\mathcal{A}_2|\mathcal{A}_1 \vee \mathcal{C}) + H(\mathcal{A}_1|\mathcal{C})$$

for sub-σ-algebras $\mathcal{A}_1$, $\mathcal{A}_2$ and $\mathcal{C}$. Evidently, $H(\mathcal{A}_1|\mathcal{C}) = H(\mathcal{A}_1 \vee \mathcal{A}_2|\mathcal{C})$ when $\mathcal{A}_2 \subset \mathcal{C}$ and $H(\mathcal{A}|\mathcal{C}) = 0$ iff $\mathcal{A} \subset \mathcal{C}$. When $\mathcal{A} \supset \mathcal{C}_1 \supset \mathcal{C}_2$,

$$I(\mathcal{A}|\mathcal{C}_1) \leq I(\mathcal{A}|\mathcal{C}_2) \quad \text{since}$$

$$I(\mathcal{A}|\mathcal{C}_2) = I(\mathcal{A} \vee \mathcal{C}_1|\mathcal{C}_2) = I(\mathcal{C}_1|\mathcal{C}_2) + I(\mathcal{A}|\mathcal{C}_1).$$

If T is an endomorphism of $(X, \mathcal{B}, m)$ and α is a countable partition with $H(\alpha) < \infty$, then $h(T, \alpha) = H(\alpha | \bigvee_{i=1}^{\infty} T^{-i}\alpha)$ is called the entropy of T with respect to α. By the Shannon-McMillan-Breiman theorem,

$$\lim_{n \to \infty} \frac{1}{n} H(\bigvee_{i=0}^{n-1} T^{-i}\alpha) = h(T, \alpha) \qquad (\text{when } H(\alpha) < \infty).$$

The entropy of T is defined by

$$h(T) = \sup\{h(T, \alpha) : \alpha \text{ is a countable partition with } H(\alpha) < \infty\}.$$

Since we are working with Lebesgue spaces,

$$h(T) = \sup\{h(T, \alpha) : \alpha \text{ is a finite partition}\}.$$

If α_n are such that $H(\alpha_n) < \infty$ and $\alpha_n \uparrow \mathcal{B}$ then $h(T) = \lim_{n \to \infty} h(T, \alpha_n)$. If α (with $H(\alpha) < \infty$) is a strong generator or, when T is an automorphism, a generator, then $h(T) = h(T, \alpha)$. The last two results are the main practical tools for the calculation of entropy.

If ϕ is a homomorphism from $(X_1, \mathcal{B}_1, m_1)$ to $(X_2, \mathcal{B}_2, m_2)$ then $I(\alpha | \mathcal{C}) \circ \phi = I(\phi^{-1}\alpha | \phi^{-1}\mathcal{C})$ whenever α is a partition of X_2 and $\mathcal{C} \subset \mathcal{B}_2$ is a σ-algebra. It follows that $H(\alpha | \mathcal{C}) = H(\phi^{-1}\alpha | \phi^{-1}\mathcal{C})$ and that $h(T_1) \geq h(T_2)$ when T_2 is a factor of T_1. Thus, $h(T_1) = h(T_2)$ when T_1 and T_2 are isomorphic.

Having established entropy as an isomorphism invariant, we conclude the section by listing some results as exercises.

Exercise. Let $(X, \mathcal{B}, m)$ be a Lebesgue space and let $\mathcal{A} \subset \mathcal{B}$ be a σ-algebra. Show that there exists a sequence of finite partitions α_n with $\alpha_n \uparrow \mathcal{A}$.

Exercise. Let T be an endomorphism of $(X, \mathcal{B}, m)$. Show that $h(T^n) = n h(T)$ for $n = 0, 1, 2, \ldots$ and that $h(T^{-1}) = h(T)$ when T is invertible.

Exercise. For $i = 1, 2$, let T_i be an endomorphism of $(X_i, \mathcal{B}_i, m_i)$. Show that $h(T_1 \times T_2) = h(T_1) + h(T_2)$.

Exercise. If T is the Bernoulli automorphism given by the probability vector $(p(1), \ldots, p(n))$, show that $h(T) = -\sum_{i=1}^{n} p(i) \log p(i)$.

Exercise. If T is the Markov automorphism given by the irreducible stochastic matrix P with invariant probability vector p $(pP = p)$, show that $h(T) = -\sum_{i,j} p(i)P(i, j) \log P(i, j)$.

Exercise. Let α, β be finite partitions of $(X, \mathcal{B}, m)$. Show that $H(\alpha \vee \beta) = H(\alpha) + H(\beta)$ iff α and β are independent (i.e. iff $m(A \cap B) = m(A) \cdot m(B)$ for all $A \in \alpha$, $B \in \beta$).

Exercise. Let α, β, γ be finite partitions of $(X, \mathcal{B}, m)$. For $B \in \beta$ denote by (B, m_B) the normalization of the restriction of m to B with $m_B(E) = \frac{m(E \cap B)}{m(B)}$. By using the last exercise on each of (B, m_B), show that $H(\alpha|\beta \vee \gamma) = H(\alpha|\beta)$ iff $\frac{m(A \cap B \cap C)}{m(B \cap C)} = \frac{m(A \cap B)}{m(B)}$ for all $A \in \alpha$, $B \in \beta$, $C \in \gamma$.

7. TYPES OF CLASSIFICATION

As we have already indicated, these notes are mainly concerned with various classifications of processes (in particular, Markov chains) and of topological Markov chains. This section is intended as a preview of the basic definitions of these classifications.

Let $(X_i, \mathcal{B}_i, \mathcal{A}_i, m_i, T_i)$, where $T_i^{-1}\mathcal{A}_i \subset \mathcal{A}_i \subset \mathcal{B}_i$ and $T_i^n \mathcal{A}_i \uparrow \mathcal{B}_i$, be processes $(i = 1, 2)$. An isomorphism $T_1 \xrightarrow{\phi} T_2$ is said to be regular if $\phi^{-1}\mathcal{A}_2 \subset T_1^p \mathcal{A}_1$ and $\phi\mathcal{A}_1 \subset T_2^p \mathcal{A}_2$ for some integer $p \geq 0$. The idea is that the code ϕ (and its inverse) should depend, perhaps on the entire past but, only on a bounded amount of the future. The main result for regular isomorphisms is that the information functions $I(\mathcal{A}_1|T_1^{-1}\mathcal{A}_1)$ and $I(\mathcal{A}_2|T_2^{-1}\mathcal{A}_2) \circ \phi$ are related by an equation. We exploit this equation in various ways to obtain invariants. The same equation holds, in a slightly weaker form, for quasi-regular isomorphisms. Quasi-regular isomorphisms are defined by insisting that the pasts of the processes should not be too distant from each other in a sense we shall make precise later.

If $(X, \mathcal{B}, m, T)$ and $(X', \mathcal{B}', m', T')$, $X = \prod_{-\infty}^{\infty} X_0$ and $X' = \prod_{-\infty}^{\infty} X_0'$, are countable state processes, a homomorphism $T \xrightarrow{\phi} T'$ is completely determined by a function $\phi_0 : X \to X_0'$ $(\phi(x) = \{\phi_0(T^n x)\})$. For a point $x \in X$ the present, $\phi_0(x)$, may depend on the entire past and future of x. If we require for each $x \in X$ that $\phi_0(x)$ is determined by a finite section of x, we have a finitary homomorphism. For a finitary homomorphism ϕ, we may define the code-length $l : X \to \mathbb{N}$ where $l(x)$ indicates how far into the past and future we need to go to determine $\phi_0(x)$; ϕ is said to have finite expected code-length if $\int l \, dm < \infty$. An isomorphism is finitary if both ϕ and ϕ^{-1} are. Keane and Smorodinsky have improved Ornstein's result by showing that entropy is a complete invariant for finitary isomorphism of Bernoulli processes. Chapter III is devoted to this result and to an information obstruction to its refinement to finitary isomorphisms with finite expected code-lengths.

A block-code is a finitary homomorphism between two finite state processes for which the present depends on a bounded amount of the past and the future (i.e. for which the code-length l is a bounded function). A block code may be regarded as a continuous homomorphism. The block isomorphism problem is, thus, simultaneously measure-theoretic and topological.

The classification of topological Markov chains with respect to topological conjugacy is discussed in Chapter V. The basic invariant, topological entropy, is far from complete. In fact there are examples which are not topologically conjugate to their inverses. There is also a problem (Williams's problem) in formulating topological conjugacy as a manageable algebraic equivalence relation of defining matrices. A more satisfactory picture emerges when we consider weaker equivalence relations: topological entropy is a complete invariant for finite equivalence, and topological entropy and period together completely characterize almost topological conjugacy.

CHAPTER II: THE INFORMATION COCYCLE

1. REGULAR ISOMORPHISMS

1. Definition. Two endomorphisms S_1, S_2 are said to be shift-equivalent if there exist homomorphisms $S_1 \xrightarrow{\phi} S_2$ and $S_2 \xrightarrow{\psi} S_1$ satisfying $\psi \circ \phi = S_1^p$, $\phi \circ \psi = S_2^p$, for some positive integer p.

2. Definition. A regular isomorphism between two processes $(X_i, \mathcal{B}_i, \mathcal{A}_i, m_i, T_i)$ $(i = 1, 2)$ is an isomorphism $T_1 \xrightarrow{\phi} T_2$ such that $\varphi^{-1}\mathcal{A}_2 \subset T_1^p\mathcal{A}_1$, $\varphi\mathcal{A}_1 \subset T_2^p\mathcal{A}_2$ for some positive integer p.

The basic result connecting regular isomorphism and shift-equivalence is

3. Proposition [F. P.]. Let $(X_i, \mathcal{B}_i, \mathcal{A}_i, m_i, T_i)$ be processes and let $S_i : X_i' \to X_i'$ be the factor endomorphism of T_i with respect to $\mathcal{A}_i$ $(i = 1, 2)$. T_1, T_2 are regularly isomorphic iff S_1, S_2 are shift-equivalent.

Proof. Let $T_1 \xrightarrow{\pi_1} S_1$, $T_2 \xrightarrow{\pi_2} S_2$ be the factor homomorphisms. Suppose T_1, T_2 are regularly isomorphic i. e. suppose there is $T_1 \xrightarrow{\rho} T_2$ and $p > 0$ with $T_1^p\mathcal{A}_1 \supset \rho^{-1}\mathcal{A}_2$ and $T_2^p\mathcal{A}_2 \supset \rho\mathcal{A}_1$. Put $\varphi = \rho T_1^p$, $\psi = \rho^{-1}T_2^p$. Then $\varphi^{-1}\mathcal{A}_2 \subset \mathcal{A}_1$ and, considering the σ-algebraic homomorphism $\phi^{-1} : \mathcal{A}_2 \to \mathcal{A}_1$ and using the fact that the spaces are Lebesgue, we obtain a homomorphism $\varphi' : X_1' \to X_2'$ with $\pi_2\varphi = \varphi'\pi_1$ and $\phi' S_1 = S_2\varphi'$. Similarly $\psi^{-1}\mathcal{A}_1 \subset \mathcal{A}_2$ and we obtain $\psi' : X_2' \to X_1'$ with $\pi_1\psi = \psi'\pi_2$ and $\psi' S_2 = S_1\psi'$. Since $\psi \circ \varphi = T_1^{2p}$ and $\phi \circ \psi = T_2^{2p}$, it follows that $\psi' \circ \phi' = S_1^{2p}$, $\phi' \circ \psi' = S_2^{2p}$ and S_1, S_2 are shift-equivalent.

For the converse, suppose $S_1 \xrightarrow{\phi'} S_2$, $S_2 \xrightarrow{\psi'} S_1$ satisfy $\psi' \circ \phi' = S_1^p$, $\phi' \circ \psi' = S_2^p$. Extend ϕ', ψ' to homomorphisms $T_1 \xrightarrow{\phi} T_2$, $T_2 \xrightarrow{\psi} T_1$. (See the following exercise 4.) Now $\psi \circ \phi$ extends $\psi' \circ \varphi' = S_1^p$ and, by the uniqueness in 4, $\psi \circ \phi = T_1^p$. Similarly, $\phi \circ \psi = T_2^p$. Since T_1, T_2 are automorphisms, φ, ψ are isomorphisms. Moreover, $\phi\mathcal{A}_1 = T_2^p\psi^{-1}\mathcal{A}_1 \subset T_2^p\mathcal{A}_2$ and $\phi^{-1}\mathcal{A}_2 \subset \mathcal{A}_1 \subset T_1^p\mathcal{A}_1$. Hence T_1, T_2 are regularly isomorphic. //

4. <u>Exercise.</u> Let T_i be automorphisms of the Lebesgue spaces $(X_i, \mathcal{B}_i, m_i)$ with exhaustive sub-σ-algebras $\mathcal{A}_i$ $(i = 1, 2)$. Suppose $U : \mathcal{A}_2 \to \mathcal{A}_1$ is a σ-algebraic homomorphism such that $T_1^{-1}U(A) = UT_2^{-1}(A)$ for all $A \in \mathcal{A}_2$. Show that U can be uniquely extended to a σ-algebraic homomorphism $\overline{U} : \mathcal{B}_2 \to \mathcal{B}_1$ such that $T_1^{-1}\overline{U}(B) = \overline{U}T_2^{-1}(B)$ for all $B \in \mathcal{B}_2$. Deduce that there is a unique homomorphism $T_1 \xrightarrow{\phi} T_2$ which extends U in the sense that $\phi^{-1}(A) = U(A)$ for all $A \in \mathcal{A}_2$.

5. <u>Definition.</u> Let $(X, \mathcal{B}, m)$ be a probability space with an endomorphism T. A <u>coboundary</u> is a function of the form $f \circ T - f$, where f is a real valued function. $f \circ T - f$ is an <u>L^q-coboundary</u> if $f \in L^q$ $(1 \le q \le \infty)$. Two real valued functions are <u>cohomologous</u> if they differ by a coboundary and <u>L^q-cohomologous</u> if they are in L^q and they differ by an L^q-coboundary.

6. <u>Definition.</u> Let $(X, \mathcal{B}, \mathcal{A}, m, T)$ be a process. $I_T = I(\mathcal{A} | T^{-1}\mathcal{A})$ is called the <u>information cocycle</u> of the process.

The following theorem provides the main tool for obtaining invariants of regular isomorphism.

7. <u>Theorem.</u> <u>Let ϕ be a homomorphism from the process</u> $(X_1, \mathcal{B}_1, \mathcal{A}_1, m_1, T_1)$ <u>to the process</u> $(X_2, \mathcal{B}_2, \mathcal{A}_2, m_2, T_2)$ <u>such that $\phi^{-1}\mathcal{A}_2 \subset T_1^{p}\mathcal{A}_1$. Let $1 \le q \le \infty$. If I_{T_1}, $I(\mathcal{A}_1 | T_1^{-p}\phi^{-1}\mathcal{A}_2) \in L^q(X_1)$ (resp. are finite a.e.) then $I_{T_2} \in L^q(X_2)$ (resp. is finite a.e.) and I_{T_1}, $I_{T_2} \circ \phi$ are L^q-cohomologous (resp. are cohomologous). When ϕ is a regular isomorphism then $I_{T_1} \in L^q(X_1)$ (resp. is finite a.e.) if and only iff $I_{T_2} \in L^q(X_2)$ (resp. is finite a.e.) and, in this case, I_{T_1} and $I_{T_2} \circ \phi$ are L^q-cohomologous (resp. are cohomologous).</u>

<u>Proof.</u> First note that if ϕ is a regular isomorphism then $\mathcal{A}_1 \supset T_1^{-p}\phi^{-1}\mathcal{A}_2 \supset T_1^{-2p}\mathcal{A}_1$ so that (see section 6 of Chapter I)

$$I(\mathcal{A}_1 | T_1^{-p}\phi^{-1}\mathcal{A}_2) \le I(\mathcal{A}_1 | T_1^{-2p}\mathcal{A}_1) = I_{T_1} + I_{T_1} \circ T_1 + \dots + I_{T_1} \circ T_1^{2p-1}$$

and $I(\mathcal{A}_1 | T_1^{-p}\phi^{-1}\mathcal{A}_2)$ is in the same Lebesgue class as I_{T_1}. Hence it is sufficient to prove the first part of the theorem.

We have

$$I(\mathcal{A}_1 | T_1^{-p-1}\phi^{-1}\mathcal{A}_2) = I(\mathcal{A}_1 \vee T_1^{-1}\mathcal{A}_1 | T_1^{-p-1}\phi^{-1}\mathcal{A}_2)$$

$$= I(T_1^{-1}\mathcal{A}_1 | T_1^{-p-1}\phi^{-1}\mathcal{A}_2) + I(\mathcal{A}_1 | T_1^{-1}\mathcal{A}_1) \text{ and}$$

$$I(\mathcal{A}_1 | T_1^{-p-1}\phi^{-1}\mathcal{A}_2) = I(\mathcal{A}_1 \vee T_1^{-p}\phi^{-1}\mathcal{A}_2 | T_1^{-p-1}\phi^{-1}\mathcal{A}_2)$$

$$= I(T_1^{-p}\phi^{-1}\mathcal{A}_2 | T_1^{-p-1}\phi^{-1}\mathcal{A}_2) + I(\mathcal{A}_1 | T_1^{-p}\phi^{-1}\mathcal{A}_2)$$

so that $I_{T_1} + g \circ T_1 = g + I_{T_2} \circ \phi \circ T_1^p$ where $g = I(\mathcal{A}_1 | T_1^{-p}\phi^{-1}\mathcal{A}_2)$. Hence I_{T_2} is finite (resp. in L^q) if I_{T_1}, $I(\mathcal{A}_1 | T_1^{-p}\phi^{-1}\mathcal{A}_2)$ are finite (resp. in L^q). Since $I_{T_2} \circ \phi \circ T_1^p$ and $I_{T_2} \circ \phi$ are cohomologous (L^q-cohomologous if $I_{T_2} \in L^q$), the theorem follows. //

8. Corollary. Let T_i be endomorphisms of $(X_i, \mathcal{B}_i, m_i)$ $(i = 1, 2)$ and let $T_1 \xrightarrow{\phi} T_2$. If $I(\mathcal{B}_1 | T_1^{-1}\mathcal{B}_1)$, $I(\mathcal{B}_1 | \phi^{-1}\mathcal{B}_2) \in L^q(X_1)$ (resp. are finite a.e.) then $I(\mathcal{B}_2 | T_2^{-1}\mathcal{B}_2) \in L^q(X_2)$ (resp. is finite a.e.) and $I(\mathcal{B}_1 | T_1^{-1}\mathcal{B}_1)$, $I(\mathcal{B}_2 | T_2^{-1}\mathcal{B}_2) \circ \phi$ are L^q-cohomologous (resp. are cohomologous).

Proof. Consider the natural extensions of T_1, T_2 and apply the theorem (with $p = 0$). //

If T is an endomorphism of $(X, \mathcal{B}, m)$, we define a linear isometry $U_T : L^2(X) \to L^2(X)$ by $f \mapsto f \circ T$. U_T is unitary if T is an automorphism. If, in 7, ϕ is a regular isomorphism and one of the information cocycles is in L^2 then we have a cocycle-coboundary equation $f = f' + U_{T_1}(g) - g$ where $f, f', g \in L^2(X_1)$. Thus, the analysis of the equation may be treated as a Hilbert space problem.

2. UNITARY OPERATORS AND COCYCLES

Throughout this section H will be a Hilbert space and U a unitary operator on H. We wish to investigate cocycle-coboundary equations $u = v + Uw - w$ where $u, v, w \in H$.

9. Definition. $u, v \in H$ are said to be cohomologous with respect to U if $u = v + Uw - w$ for some $w \in H$. $u \in H$ is called a coboundary if it is cohomologous to 0.

For $x \in H$ we put $S_n(x) = x + Ux + \dots + U^{n-1}x$ and $\sigma^2(x) = \lim_{n\to\infty} \frac{1}{n} \|S_n(x)\|^2$ when this limit exists.

10. Proposition [F. P.]. Let $u, v \in H$ be cohomologous (with respect to U). If $\sigma^2(u)$ exists then $\sigma^2(v)$ also exists and $\sigma^2(u) = \sigma^2(v)$.

Proof. Denote by P the orthogonal projection of H onto the U-fixed vectors, $\{x \in H : x = Ux\}$. Suppose $u = v + Uw - w$, $w \in H$. Successively applying U to this equation and summing,

$$S_n(u) = S_n(v) + U^n w - w.$$

Hence

$$\|S_n(u)\|^2 = \|S_n(v)\|^2 + \|U^n w - w\|^2 + 2\mathcal{R}e\langle S_n(v), U^n w - w\rangle$$
$$= \|S_n(v)\|^2 + \|U^n w - w\|^2 + 2\mathcal{R}e\langle U^{-1}v + \ldots + U^{-n}v, w\rangle - 2\mathcal{R}e\langle v + \ldots + U^{n-1}v, w\rangle.$$

The result follows by noting that $\frac{1}{n}\|U^n w - w\|^2 \le \frac{4}{n}\|w\|^2 \to 0$ and that, by von Neumann's ergodic theorem, $\frac{1}{n}(U^{-1}v + \ldots + U^{-n}v) \to P(v)$ and $\frac{1}{n}(v + \ldots + U^{n-1}v) \to P(v)$. //

11. Proposition. $x \in H$ is a coboundary iff the sequence $\|x + Ux + \ldots + U^n x\|$ $(n = 0, 1, \ldots)$ is bounded.

Proof. If $x = Uy - y$, then $\|x + \ldots + U^n x\| = \|U^{n+1}y - y\| \le 2\|y\|$ and the sequence is bounded. Now suppose there exists $k \in \mathbb{R}$ such that $\|x + Ux + \ldots + U^n x\| \le k$ for $n = 0, 1, \ldots$. Let S be the set of all convex combinations of $\{S_n(x)\}$,

$$S = \{\lambda_0 x + \lambda_1 S_1(x) + \ldots + \lambda_n S_n(x) : \lambda_i \ge 0, \sum_{i=0}^{n} \lambda_i = 1, n = 0, 1, \ldots\}.$$

Let $\bar{S}$ be the closure of S in the weak topology of H. $\bar{S}$ is convex and, as it is contained in the weakly compact set $\{z \in H : \|z\| \le k\}$, it is weakly compact. Moreover, $\bar{S}$ is invariant under the continuous affine map $z \mapsto x + Uz$. By the Schauder-Tychonov theorem this map has a fixed point in $\bar{S}$ i.e. there exists $y \in \bar{S}$ with $y = x + Uy$, and x is a coboundary. //

We now need some standard facts from the spectral theory of unitary operators, which we list here. We use the notation of the Appendix of [P. 1] where the details may be found. ($U : H \to H$ is a unitary operator.)

(i) For $x \in H$ we denote by $Z(x)$ the closure of the linear span of $\{U^n x : n \in \mathbb{Z}\}$. $Z(x)$ is called the cyclic subspace generated by x. Both $Z(x)$ and its orthogonal complement $Z(x)^{\perp}$ are invariant under U.

(ii) For each $x \in H$ there exists a unique finite positive Borel measure $\tilde{x}$ on the circle $K = \{\lambda \in \mathbb{C} : |\lambda| = 1\}$ such that

$$\langle U^n x, x\rangle = \int_K \lambda^n \, d\tilde{x}(\lambda) \quad \text{for all } n \in \mathbb{Z} \quad \text{(Herglotz's theorem)}.$$

$\tilde{x}$ is called the spectral measure of x.

(iii) If $y \in Z(x)$ then $\tilde{y} \ll \tilde{x}$, with $\tilde{y} \sim \tilde{x}$ iff $Z(y) = Z(x)$. For each finite positive Borel measure $\mu \ll \tilde{x}$ there exists $y \in Z(x)$ with $\tilde{y} = \mu$. If $y, z \in Z(x)$ then $Z(y) \perp Z(x)$ iff $\tilde{y} \perp \tilde{z}$.

(iv) $U|Z(x)$ is unitarily equivalent to M on $L^2(\tilde{x})$ where the multiplication operator M is defined $(Mf)(\lambda) = \lambda f(\lambda)$, $f \in L^2(\tilde{x})$. The Hilbert space isomorphism (i.e. bijective linear isometry) effecting this equivalence is obtained by extending the map $U^n x \mapsto \lambda^n$, $n \in \mathbb{Z}$. In particular, x corresponds to the constant function 1. If $y \in Z(x)$, y corresponds to a function $\alpha(\lambda)\left(\frac{d\tilde{y}}{d\tilde{x}}\right)^{\frac{1}{2}}$, where $|\alpha| = 1$.

(v) If $u, v \in H$ and $\tilde{u} = \tilde{v}$ then there exists a unitary operator $W : H \to H$ such that $WU = UW$ and $Wu = v$. W is obtained by composing the two maps conjugating $U|_{Z(u)}$ and $U|_{Z(v)}$ to M on $L^2(\tilde{u})$, and extending.

12. Theorem [P. S.]. Suppose the unitary operator $U : H \to H$ has no fixed vectors other than 0 and let $u, v \in H$. The following are equivalent:

(a) There exists a unitary operator $W : H \to H$ such that $WU = UW$ and $u = Wv + Uw - w$ for some $w \in H$.

(b) For some measure (and therefore all measures) ρ on the circle K satisfying $\rho \gg \tilde{u}, \tilde{v}$ and $\rho(1) = 0$ (i.e. the point 1 is not an atom),

$$\int_K \left[\left(\frac{d\tilde{u}}{d\rho}\right)^{\frac{1}{2}} - \left(\frac{d\tilde{v}}{d\rho}\right)^{\frac{1}{2}}\right]^2 \frac{1}{|\lambda-1|^2} \, d\rho < \infty.$$

13. Remarks. (i) For ρ satisfying $\rho \gg \tilde{u}, \tilde{v}$ and $\rho(1) = 0$

$$J(\rho) = \int_K \left[\left(\frac{d\tilde{u}}{d\rho}\right)^{\frac{1}{2}} - \left(\frac{d\tilde{v}}{d\rho}\right)^{\frac{1}{2}}\right]^2 \frac{1}{|\lambda-1|^2} \, d\rho$$ is independent of ρ: If $\sigma \gg \rho \gg \tilde{u}, \tilde{v}$ and $\sigma(1) = \rho(1) = 0$, it is easy to see that $J(\sigma) = J(\rho)$. For arbitrary ρ_1, ρ_2 with $\rho_1, \rho_2 \gg \tilde{u}, \tilde{v}$ and $\rho_1(1) = \rho_2(1) = 0$, $\rho_1 + \rho_2 \gg \rho_1, \rho_2$ and so $J(\rho_1) = J(\rho_1 + \rho_2) = J(\rho_2)$.

(ii) The condition that U has no fixed vectors other than 0 guarantees that $\tilde{w}(1) = 0$ for all $w \in H$. More generally, if $\tilde{w}(\alpha) > 0$ for some $w \in H$ and $\alpha \in K$ then $\chi_{\{\alpha\}} \in L^2(\tilde{w})$ is non-trivial, $M(\chi_{\{\alpha\}}) = \alpha \chi_{\{\alpha\}}$ and, taking preimages with respect to the canonical map of $Z(w)$ onto $L^2(\tilde{w})$, we see that α is an eigenvalue of U.

Proof of 12. Suppose (a) holds. Since $\widetilde{Wv} = \tilde{v}$, to deduce (b) we may assume $u = v + Uw - w$. Write $u = u' + x$, $v = v' + y$ with $x, y \in Z(w)$ and $u', v' \perp Z(w)$. Then $x = y + Uw - w$, $u' = v'$ and $\tilde{u} = \tilde{u}' + \tilde{x}$, $\tilde{v} = \tilde{v}' + \tilde{y}$. The relation $x = y + Uw - w$ implies $f(\lambda) = g(\lambda) + \lambda - 1$ where $f = \left(\frac{d\tilde{x}}{d\tilde{w}}\right)^{\frac{1}{2}} \alpha(\lambda)$, $g = \left(\frac{d\tilde{y}}{d\tilde{w}}\right)^{\frac{1}{2}} \beta(\lambda)$, $|\alpha| = |\beta| = 1$. Now for $\rho \gg \tilde{u}, \tilde{v}$, $\rho(1) = 0$ we have

$$\begin{aligned}
J(\rho) &= \int_K \left[\left(\frac{d(\tilde{u}' + \tilde{x})}{d\rho}\right)^{\frac{1}{2}} - \left(\frac{d(\tilde{v}' + \tilde{y})}{d\rho}\right)^{\frac{1}{2}}\right]^2 \frac{1}{|\lambda-1|^2}\, d\rho \\
&\le \int_K \left[\left(\frac{d\tilde{x}}{d\rho}\right)^{\frac{1}{2}} - \left(\frac{d\tilde{y}}{d\rho}\right)^{\frac{1}{2}}\right]^2 \frac{1}{|\lambda-1|^2}\, d\rho \quad \left(\text{since } \left|(a+b)^{\frac{1}{2}} - (c+b)^{\frac{1}{2}}\right| \le \left|a^{\frac{1}{2}} - c^{\frac{1}{2}}\right|\right) \\
&= \int_K \left[\left(\frac{d\tilde{x}}{d\tilde{w}}\right)^{\frac{1}{2}} - \left(\frac{d\tilde{y}}{d\tilde{w}}\right)^{\frac{1}{2}}\right]^2 \frac{1}{|\lambda-1|^2}\, d\tilde{w} \\
&= \int_K (|f| - |g|)^2 \frac{1}{|\lambda-1|^2}\, d\tilde{w} \\
&\le \int_K \frac{|f-g|^2}{|\lambda-1|^2}\, d\tilde{w} = \tilde{w}(K) < \infty .
\end{aligned}$$

Now suppose (b) holds. Choose $w \in H$ such that $\tilde{w} \gg \tilde{u}, \tilde{v}$. (E.g. write $\tilde{u} = \mu_a + \mu_s$ with $\mu_a \ll \tilde{v}$, $\mu_s \perp \tilde{v}$ and let $z \in Z(u)$ have $\tilde{z} = \mu_s$. Now $\tilde{z} \perp \tilde{v}$ so $\widetilde{z+v} = \tilde{z} + \tilde{v}$. Take $w = z + v$.) Let u_0, v_0 be the preimages, under the canonical map of $Z(w)$ to $L^2(\tilde{w})$, of $\left(\frac{d\tilde{u}}{d\tilde{w}}\right)^{\frac{1}{2}}$, $\left(\frac{d\tilde{v}}{d\tilde{w}}\right)^{\frac{1}{2}}$ respectively. Then $\tilde{u}_0 = \tilde{u}$, $\tilde{v}_0 = \tilde{v}$ and there exist unitary $R, S : H \to H$ such that $RU = UR$, $SU = US$ and $Ru = u_0$, $Sv = v_0$. By hypothesis, $h(\lambda) = \left[\left(\frac{d\tilde{u}}{d\tilde{w}}\right)^{\frac{1}{2}} - \left(\frac{d\tilde{v}}{d\tilde{w}}\right)^{\frac{1}{2}}\right] \frac{1}{(\lambda-1)} \in L^2(\tilde{w})$. Writing this as

$$\left(\frac{d\tilde{u}}{d\tilde{w}}\right)^{\frac{1}{2}} = \left(\frac{d\tilde{v}}{d\tilde{w}}\right)^{\frac{1}{2}} + \lambda h(\lambda) - h(\lambda)$$

and taking preimages with respect to the canonical map of $Z(w)$ onto $L^2(\tilde{w})$ we obtain $u_0 = v_0 + Uw_0 - w_0$ for some $w_0 \in Z(w)$. Hence $Ru = Sv + Uw_0 - w_0$ and (a) follows with $W = R^{-1}S$. //

A partial converse to 10 is

14. Proposition. Assume U has no fixed vectors other than 0, let $u, v \in H$ and denote by μ the Lebesgue measure on K. Suppose $\tilde{u}, \tilde{v} \ll \mu$, $\left(\frac{d\tilde{u}}{d\mu}\right)^{\frac{1}{2}}$ and $\left(\frac{d\tilde{v}}{d\mu}\right)^{\frac{1}{2}}$ are differentiable at 1 and $\sigma^2(u) = \sigma^2(v)$. Then there exists a unitary

$W : H \to H$ such that $WU = UW$ and $u = Wv + Uw - w$ for some $w \in H$.

Proof. $\sigma^2(u) = \lim_{n\to\infty} \frac{1}{n} \|u + Uu + \ldots + U^{n-1}u\|^2$

$$= \lim_{n\to\infty} \frac{1}{n} \int_K |1 + \lambda + \ldots + \lambda^{n-1}|^2 \, d\tilde{u}$$

$$= \lim_{n\to\infty} \frac{1}{n} \int_K \left| \frac{1-\lambda^n}{1-\lambda} \right|^2 \cdot \frac{d\tilde{u}}{d\mu} \, d\mu$$

$$= \frac{d\tilde{u}}{d\mu}(1), \text{ by a well known property of Féjer's kernel.}$$

By hypothesis, then, $\frac{d\tilde{u}}{d\mu}(1) = \frac{d\tilde{v}}{d\mu}(1)$. Hence

$$J(\mu) = \int_K \left[\frac{\left(\frac{d\tilde{u}}{d\mu}\right)^{\frac{1}{2}} - \left(\frac{d\tilde{u}}{d\mu}(1)\right)^{\frac{1}{2}}}{1-\lambda} - \frac{\left(\frac{d\tilde{v}}{d\mu}\right)^{\frac{1}{2}} - \left(\frac{d\tilde{v}}{d\mu}(1)\right)^{\frac{1}{2}}}{1-\lambda} \right]^2 d\mu < \infty$$

and the proposition follows from 12. //

15. Corollary. Suppose U has no fixed vectors other than 0 and let $u \in H$. If $\tilde{u} \ll \mu$ and if $\left(\frac{d\tilde{u}}{d\mu}\right)^{\frac{1}{2}}$ is differentiable at 1 then, $\sigma^2(u) = 0$ iff u is a coboundary.

Proof. If $\sigma^2(u) = 0$, taking $v = 0$ in 14, $u = Uw - w$ for some $w \in H$. The converse is obvious. //

3. INFORMATION VARIANCE

16. Definition. Let $(X, \mathcal{B}, \mathcal{A}, m, T)$ be a process with information cocycle $I_T = I(\mathcal{A} | T^{-1}\mathcal{A})$. $I_T^0 = I_T - \int I_T \, dm$ is called the centralised information cocycle of the process (when $I_T \in L^1(X)$). If $I_T \in L^2(X)$ and $\sigma^2(I_T^0)$ exists, then $\sigma^2(T) = \sigma^2(I_T^0)$ is called the information variance of the process.

By 7, when $T_1 \overset{\phi}{\to} T_2$ is a regular isomorphism of the processes $(X_i, \mathcal{B}_i, \mathcal{A}_i, m_i, T_i)$ with $I_{T_i} \in L^2(X_i)$ $(i = 1, 2)$ we have

$$I_{T_1} = I_{T_2} \circ \phi + g \circ T_1 - g \quad \text{for some } g \in L^2(X_1).$$

From this we obtain $\int I_{T_1} \, dm_1 = \int I_{T_2} \, dm_2$ and

$$I_{T_1}^0 = I_{T_2}^0 \circ \phi + g \circ T_1 - g.$$

Now 10 shows that $\sigma^2(T_1)$ exists iff $\sigma^2(T_2)$ exists and that in this case $\sigma^2(T_1) = \sigma^2(T_2)$. Thus information variance is, under suitable conditions, an invariant of regular isomorphism. We now turn to its computation.

17. Exercise [F. P.]. Let S and T be processes. Assuming these quantities exist, prove that $\sigma^2(T^n) = n\sigma^2(T)$, $n = 1, 2, \ldots$ and that $\sigma^2(S \times T) = \sigma^2(S) + \sigma^2(T)$.

18. Exercise. Let T be the Bernoulli process given by the probability vector $p = (p(1), \ldots, p(k))$ and let α be its state partition. Show that $I_T = I(\alpha)$ and, using the independence of $\{I(\alpha) \circ T^n\}$, that

$$\sigma^2(T) = \int (I(\alpha) - h)^2 = \sum_{i=1}^{k} (\log p(i) + h)^2 p(i) \quad \text{where } h = -\sum_{i=1}^{k} p(i) \log p(i).$$

Let $(X, \mathcal{B}, m, T)$ be the Markov chain given by the matrix P with left invariant probability vector p, $pP = p$. It is easy to see that $I_T = I(\alpha | T^{-1}\alpha)$, where α is the state partition.

$$\text{i.e. } I_T(x) = \log\left(\frac{p(x_1)}{p(x_0) P(x_0, x_1)}\right) \quad \text{for } x = (x_n) \in X.$$

Hence I_T and I_T^0 are both functions of only two coordinates and $\sigma^2(T)$ may be evaluated via

19. Proposition [P. S.]. Let $(X, \mathcal{B}, m, T)$ be the Markov chain given by the aperiodic $k \times k$ matrix P and the probability vector p with $pP = p$. Consider $L^2(X)$ with the unitary operator U induced by T, $Uf = f \circ T$ for $f \in L^2(X)$. If $u \in L^2(X)$ is a function of two variables with $\int u\,dm = 0$ then it has spectral measure $\tilde{u} \ll \mu$ (Lebesgue measure) such that $f = \frac{d\tilde{u}}{d\mu}$ is C^∞ and

$$\sigma^2(u) = f(1) = \int |u|^2 dm + 2\mathcal{R}e \sum_{i,j=1}^{k} p(i) P(i, j) \overline{c(j)} u(i, j)$$

where $c \in V_0 = \{v \in \mathbb{C}^k : P^n v \to 0 \text{ as } n \to \infty\}$ satisfies $(I - P)c = b$,

$$b(i) = \sum_{j=1}^{k} P(i, j) u(i, j).$$

Proof. First note that for $y = (y(1), \ldots, y(k))^{tr} \in \mathbb{C}^k$

$P^n y \to (\sum_{i=1}^{k} p(i) y(i)) \mathbf{1}$ where $\mathbf{1} = (1, \ldots, 1)^{tr}$, since $P^n \to \begin{pmatrix} p \\ \vdots \\ p \end{pmatrix}$, the matrix with identical rows p.

For $n \geq 1$ we have

$$\langle U^n u, u \rangle = \int u(T^n) \bar{u}\, dm$$

$$= \sum_{i,j,k,l} \overline{u(i,j)}\ u(k,l)\, p(i)\, P(i,j)\, P^{n-1}(j,k)\, P(k,l)$$
$$= (P^{n-1}b, a)$$

where $a(j) = \sum_i u(i,j)\, p(i)\, P(i,j)$, $b(k) = \sum_l u(k,l)\, P(k,l)$ and $(\cdot,\cdot)$ denotes the usual $\mathbb{C}^k$ inner product. Note $\sum_k p(k)b(k) = \sum_{k,l} p(k)u(k,l)P(k,l) = \int u\,dm = 0$ so that $b \in V_0$. Since $P^n b \to 0$ exponentially fast, $\langle U^n u, u\rangle \to 0$ exponentially fast and the function $f(\lambda) = \sum_{n\in\mathbb{Z}} \langle U^n u, u\rangle \lambda^{-n}$ is defined and C^∞ on the circle K. It is easy to see that

$$\langle U^n u, u\rangle = \int_K \lambda^n f(\lambda)\, d\mu(\lambda) \quad \text{for } n \in \mathbb{Z}.$$

Hence $\tilde{u} \ll \mu$ and $\frac{d\tilde{u}}{d\mu} = f$. That $\sigma^2(u) = f(1)$ was shown in the proof of 14. To evaluate $f(\lambda)$ for $\lambda \in K$ note that, by the exponential convergence of $P^n b$ to 0, $\sum_{n=1}^{\infty} (\lambda^{-1}P)^{n-1} b$ converges, to $c(\lambda)$ say. Then $c(\lambda) \in V_0$, $(I - \lambda^{-1}P)c(\lambda) = b$ and these two conditions determine $c(\lambda)$, since $(I - \lambda^{-1}P)z = 0$ and $z \in V_0$ together imply $z = 0$. Now for $\lambda \in K$

$$f(\lambda) = \int |u|^2 dm + 2\mathcal{R}e \sum_{n=1}^{\infty} \lambda^{-n}(P^{n-1}b, a)$$
$$= \int |u|^2 dm + 2\mathcal{R}e(\lambda^{-1} \sum_{n=1}^{\infty} ((\lambda^{-1}P)^{n-1}b, a))$$
$$= \int |u|^2 dm + 2\mathcal{R}e\, \lambda^{-1}(c(\lambda), a)$$

and the proof is completed by taking $\lambda = 1$, $c = c(1)$. //

20. Exercise. Calculate the information variance of the Markov chain defined by the matrix $\begin{pmatrix} p & q \\ q & p \end{pmatrix}$, $0 < p, q < 1$, $p + q = 1$.

21. Exercise. Use 19 to compute the information variance of the Bernoulli process given by a probability vector p. Compare with 18.

22. Exercise. Use 19 and 15 to show, with the hypotheses and notation of 19, that $\sigma^2(u) = 0$ iff u is a coboundary. In particular, $\sigma^2(T) = 0$ iff I_T is cohomologous to the constant $h(T) = \int I_T\, dm$.

It is now possible to show that many Bernoulli processes, although isomorphic (i.e. with identical entropies), are not regularly isomorphic. For instance the process given by $(\frac{1}{4}, \frac{1}{4}, \frac{1}{4}, \frac{1}{4})$ and $(\frac{1}{2}, \frac{1}{8}, \frac{1}{8}, \frac{1}{8}, \frac{1}{8})$, which were proved by Meschalkin [M'] to be isomorphic, are not regularly isomorphic as computation of their information variances will reveal. In the next section we show, quite generally, that two Bernoulli processes given by the vectors p, q are not regularly isomorphic unless q may be obtained from p by a permutation.

4. THE VARIATIONAL PRINCIPLE FOR TOPOLOGICAL MARKOV CHAINS

We shall prove the variational principle for functions depending on finitely many coordinates of a topological Markov chain. We define pressure for such functions, and use pressure to obtain invariants of regular isomorphism.

The following variational characterisation of Markov measures is essentially due to Wolfowitz [W''].

23. Lemma. Let T be the shift on $X = \prod_{-\infty}^{\infty} \{1, \ldots, k\}$ and let α be the state partition of X. If m is a T-invariant Borel probability such that with respect to it T has entropy $h(m) = H(\alpha|T^{-1}\alpha)$, then m is a Markov measure.

Proof. From $H(\alpha|\bigvee_{i=1}^{\infty} T^{-i}\alpha) = h(m) = H(\alpha|T^{-1}\alpha)$ we deduce that $H(\alpha|\bigvee_{i=1}^{n} T^{-i}\alpha) = H(\alpha|T^{-1}\alpha)$ for all $n = 1, 2, \ldots$. By the last exercise of section 6 of Chapter I,

$$\frac{m(A_{i_0} \cap T^{-1}A_{i_1})}{m(A_{i_1})} = \frac{m(A_{i_0} \cap T^{-1}A_{i_1} \cap T^{-2}A_{i_2})}{m(A_{i_1} \cap T^{-1}A_{i_2})} = \frac{m(A_{i_0} \cap T^{-1}A_{i_1} \cap T^{-2}A_{i_2} \cap T^{-3}A_{i_3})}{m(A_{i_1} \cap T^{-1}A_{i_2} \cap T^{-2}A_{i_3})} = \ldots$$

for all sequences $A_{i_0}, A_{i_1}, \ldots$ of sets in α. Now taking

$$P(i_0, i_1) = \begin{cases} 0 & \text{if } m(A_{i_0} \cap T^{-1}A_{i_1}) = 0 \\ \dfrac{m(A_{i_0} \cap T^{-1}A_{i})}{m(A_{i_0})} & \text{otherwise} \end{cases}$$

and $p(i_0) = m(A_{i_0})$ we see that $pP = p$ and that m is the Markov measure given by P. //

24. Theorem. Let $(X, \mathcal{B}, m, T)$ be a (reduced) Markov chain. Then

$$\int I_\mu \, d\mu \leq \int I_m \, d\mu$$

for all T-invariant Borel probabilities μ on the topological Markov chain (X, T). Equality holds only when $\mu = m$.

Proof. Denote by α the state partition of X and for a T-invariant Borel probability μ put $I_\mu = I_\mu(\alpha | \bigvee_{i=1}^{\infty} T^{-i}\alpha)$, $J_\mu = I_\mu(\alpha | T^{-1}\alpha)$. By 23, $\int J_\mu d\mu \geq \int I_\mu d\mu$ with equality only when μ is a Markov measure. We have

$$\int (I_m - J_\mu) \, d\mu = \int \left[-\log\left(\frac{m[x_0, x_1]\mu[x_1]}{m[x_1]\mu[x_0, x_1]} \right) \right] d\mu$$

$$\geq \sum_{\mu[x_0, x_1] > 0} \left(1 - \frac{m[x_0, x_1]\mu[x_1]}{m[x_1]\mu[x_0, x_1]} \right) \mu[x_0, x_1]$$

since $-\log y \geq 1 - y$ for $y > 0$ with equality only when $y = 1$. Hence

$$\int I_m \, d\mu - \int I_\mu d\mu \geq \int I_m \, d\mu - \int J_\mu \, d\mu \geq 1 - \sum_{\mu[x_0, x_1] > 0} \frac{m[x_0, x_1]\mu[x_1]}{m[x_1]} \geq 0 .$$

For the last part examine the conditions under which the inequalities used are equalities. //

25. Theorem (Variational Principle) [L. R.], [S'''] . Let (X, T) be a topological Markov chain and let f be a function of two coordinates, $f(x) = f(x_0, x_1)$ for $x = (x_n) \in X$. Then there is a unique T-invariant probability m such that

$$\int (I_\mu + f) \, d\mu \leq \int (I_m + f) \, dm$$

for all T-invariant Borel probabilities μ. m is Markov and is supported by X.

Proof. Suppose (X, T) is given by the $k \times k$ irreducible 0-1 matrix A. Consider the $k \times k$ matrix M defined $M(i, j) = A(i, j) e^{f(i, j)}$. By the Perron-Frobenius theorem there is a maximum eigenvalue $\beta > 0$ and a strictly positive vector $r = (r(1), \ldots, r(k))^{tr}$ such that $Mr = \beta r$. The matrix P with $P(i, j) = M(i, j) r(j) / \beta r(i)$ is stochastic. Take m to be the Markov probability defined by P. Then for (i, j) with $A(i, j) = 1$ we have

$$-\log P(i, j) = -f(i, j) + \log \beta + \log r(i) - \log r(j)$$

and, since I_m is cohomologous to the function equal to $-\log P(i, j)$ on $[i, j]$,

$$I_m + f = \log\beta + g \circ T - g$$

for some $g : X \to \mathbb{R}$. Now from 24 we see that

$$\int (I_m + f)\, dm = \log\beta = \int (I_m + f)\, d\mu \geq \int (I_\mu + f)\, d\mu$$

with equality only when $\mu = m$. //

26. Corollary [P. 2]. *Let (X, T) be a topological Markov chain. There is a unique T-invariant probability m such that $h_m(T) \geq h_\mu(T)$ for all T-invariant Borel probabilities μ. m is Markov and is supported by X.*

27. Remark. 26 is obtained by taking $f \equiv 0$ in 25. The Markov probability m of 26 is defined by the stochastic matrix

$$P(i, j) = \frac{A(i, j)\, r(j)}{\beta\, r(i)}$$

where A is the defining matrix of (X, T), β the maximum eigenvalue of A and r a strictly positive right eigenvector. Such Markov measures and their associated Markov chains are, for reasons obvious from 26, said to be of *maximal type.*

28. Corollary. *Let (X, T) be a topological Markov chain and let f be a function of finitely many coordinates. Then there exists a unique T-invariant probability m such that*

$$\int (I_\mu + f)\, d\mu \leq \int (I_m + f)\, dm$$

for all T-invariant Borel probabilities μ. m is multiple Markov.

Proof. If f depends on n coordinates, we consider the topologically conjugate topological Markov chain having as states allowable words of length $n - 1$, with $j_1 j_2 \ldots j_{n-1}$ following $i_1 i_2 \ldots i_{n-1}$ iff $j_1 = i_2, \ldots, j_{n-2} = i_{n-1}$. //

29. Definition. If f is a function of finitely many coordinates of the topological Markov chain (X, T),

$$P(f) = \sup \{ h_\mu(T) + \int f\, d\mu : \mu \text{ is a T-invariant Borel probability} \}$$

is called the *pressure of* f. $P(0)$ is called the *topological entropy* of T.

25 may now be interpreted as stating that there is a unique T-invariant probability m (which is Markov and) which satisfies $h_m(T) + \int f\,dm = P(f)$.

30. Lemma. Let $(X, \mathcal{B}, m, T)$ be a Markov chain and write $S_n(I_m) = I_m + I_m \circ T + \ldots + I_m \circ T^{n-1}$. Then for any $t \in \mathbb{R}$,

$$\lim_{n\to\infty} \frac{1}{n} \log \int \exp(tS_n(I_m))\, dm = P((t-1)I_m).$$

Proof. Suppose the Markov chain is defined by the stochastic matrix M. Then

$$S_n(I_m) = - \sum_{i_0,\ldots,i_n} \chi_{[i_0,\ldots,i_n]} \log\left(\frac{m[i_0]M(i_0,i_1)M(i_1,i_2)\ldots M(i_{n-1},i_n)}{m[i_n]}\right)$$

and $\int \exp(tS_n(I_m))\, dm$

$$= \int \sum_{i_0,\ldots,i_n} \chi_{[i_0,\ldots,i_n]} \left(\frac{m[i_0]M(i_0,i_1)\ldots M(i_{n-1},i_n)}{m[i_n]}\right)^{-t} dm$$

$$= \sum_{i_0,\ldots,i_n} (m[i_0])^{1-t}(m[i_n])^{t}(M(i_0,i_1)\ldots M(i_{n-1},i_n))^{1-t}.$$

But there exist constants k, K such that $0 < k \le (m[i_0])^{1-t}(m[i_n])^{t} \le K$ for all i_0, i_n so that

$$k \sum_{i_0,\ldots,i_n} (M(i_0,i_1)\ldots M(i_{n-1},i_n))^{1-t} \le \int \exp(tS_n(I_m))\, dm$$
$$\le K \sum_{i_0,\ldots,i_n} (M(i_0,i_1)\ldots M(i_{n-1},i_n))^{1-t}$$

and

$$\lim_{n\to\infty} \frac{1}{n} \log \int \exp(tS_n(I_m))\, dm = \lim_{n\to\infty} \frac{1}{n} \log \sum_{i_0,\ldots,i_n} (M(i_0,i_1)\ldots M(i_{n-1},i_n))^{1-t}$$

provided the last limit exists.

Denote by M^{1-t} the matrix whose (i, j) entry is $M(i,j)^{1-t}$ if $M(i,j) > 0$ and zero otherwise. Use the Perron-Frobenius theorem to find $\beta > 0$ and a strictly positive vector r such that $M^{1-t}r = \beta r$. By an argument similar to the above

$$\lim_{n\to\infty} \frac{1}{n} \log \sum_{i_0,\ldots,i_n} (M(i_0,i_1)\ldots M(i_{n-1},i_n))^{1-t}$$

$$= \lim_{n\to\infty} \frac{1}{n} \log \sum_{i_0,\ldots,i_n} M(i_0,i_1)^{1-t} \ldots M(i_{n-1},i_n)^{1-t} r(i_n)$$

$$= \lim_{n\to\infty} \frac{1}{n} \log \sum_{i_0,\ldots,i_{n-1}} M(i_0, i_1)^{1-t} \cdots M(i_{n-2}, i_{n-1})^{1-t} (\sum_{i_n} M(i_{n-1}, i_n)^{1-t} r(i_n))$$

$$= \lim_{n\to\infty} \frac{1}{n} \log \sum_{i_0,\ldots,i_{n-1}} M(i_0, i_1)^{1-t} \cdots M(i_{n-2}, i_{n-1})^{1-t} r(i_{n-1}) \beta$$

$$= \lim_{n\to\infty} \frac{1}{n} \log (\beta^n \sum_{i_0} r(i_0)) \quad \text{(by repeating the last step } n \text{ times)}$$

$$= \log \beta .$$

Finally, the proof of 25 shows that $P((t-1) I_m) = \log \beta$. //

31. <u>Theorem [T]. Let</u> $(X_1, \mathcal{B}_1, m_1, T_1)$ <u>and</u> $(X_2, \mathcal{B}_2, m_2, T_2)$ <u>be Markov chains. Suppose they are regularly isomorphic. Then</u> $P(tI_{T_1}) = P(tI_{T_2})$ <u>for all</u> $t \in \mathbb{R}$. <u>If</u> T_1, T_2 <u>are Bernoulli and are defined by the probability vectors</u> $p = (p(1), \ldots, p(k))$, $q = (q(1), \ldots, q(l))$ <u>then</u> $l = k$ <u>and</u> q <u>may be obtained from</u> p <u>by a permutation.</u>

<u>Proof.</u> Let $T_1 \overset{\phi}{\to} T_2$ be the regular isomorphism. Note that the information cocycles of Markov chains are bounded to conclude from 7 that

$$I_{T_1} = I_{T_2} \circ \phi + g \circ T_1 - g$$

where $g \in L^\infty(X_1)$. By 30, for each $t \in \mathbb{R}$,

$$P((t-1) I_{T_1}) = \lim_{n\to\infty} \frac{1}{n} \log \int \exp t(S_n(I_{T_2}) \circ \phi + g \circ T_1^n - g)\, dm_1$$

$$= \lim_{n\to\infty} \frac{1}{n} \log \int \exp tS_n(I_{T_2}) \circ \phi\, dm_1 \quad (\text{since } g \in L^\infty(X_1))$$

$$= \lim_{n\to\infty} \frac{1}{n} \log \int \exp tS_n(I_{T_2})\, dm_2 = P(t-1) I_{T_2}) .$$

This proves the first part of the theorem. If T_1, T_2 are Bernoulli processes defined by $p = (p(1), \ldots, p(k))$, $q = (q(1), \ldots, q(l))$ then by 25 and its proof we have

$$\log \sum_{i=1}^{k} p(i)^{-t} = P(tI_{T_1}) = P(tI_{T_2}) = \log \sum_{j=1}^{l} q(j)^{-t}$$

for all $t \in \mathbb{R}$. It follows that $l = k$ and that q is a permutation of p. //

We end the section with an exercise and some remarks on variational principles

32. Exercise (Ratio Variational Principle). Let f be a strictly positive function depending on only two coordinates of a topological Markov chain (X, T). Show that there exists a unique T-invariant probability m (which is Markov) such that

$$\frac{\int I_m \, dm}{\int f \, dm} \geq \frac{\int I_\mu \, d\mu}{\int f \, d\mu}$$

for all T-invariant Borel probabilities μ.

33. Remarks. (i) Theorem 25 has not been stated in its most general form. Excellent accounts of the variational principle in a setting more general than 25 may be found in [B. 1] and [W. 3]. Bowen's monograph [B. 1] also relates topological Markov chains to Axiom A diffeomorphisms.

(ii) The general definition of pressure for $f \in C(X)$ where X is a compact metric space with continuous $T : X \to X$ and the variational principle related to this are due to Ruelle [R"] and Walters [W. 2].

(iii) The maximal measure of 26 was constructed in [P. 2]. Later it was learned that Shannon had included a similar result in his earlier paper [S. W.]. We discuss these results in the Appendix. The maximal measure has been particularly fruitful in the hands of Adler and Weiss [A. W.], Sinai [S"] and Bowen [B. 1].

(iv) 32, the ratio variational principle appears in [B. R.] and [P. 4]. (See also [T].)

(v) For a more general discussion of 30, see [R"] and [T].

5. A GROUP INVARIANT [P. 3]

34. Definition. Let $(X_1, \mathcal{B}_1, m_1)$ be a Lebesgue space and let T_1 be an endomorphism on it. If $J_1 : X_1 \to \mathbb{R}$ we let $\Lambda_{T_1}(J_1)$ be the set $\{(a, b) \in \mathbb{R}^2 : F \circ T_1 = F \exp 2\pi i(a + bJ_1)$ for some $F : X_1 \to \mathbb{C}$ with $|F| = 1\}$. If T_1 is a process with finite information cocycle I_{T_1}, we put $\Lambda(T_1) = \Lambda_{T_1}(I_{T_1})$.

It is easy to see that $\Lambda_{T_1}(J_1)$ is a subgroup of $\mathbb{R}^2$. If $J_1 = J_1' + g \circ T_1 - g$ where $J_1', g : X_1 \to \mathbb{R}$ then $\Lambda_{T_1}(J_1') = \Lambda_{T_1}(J_1)$. If $(X_2, \mathcal{B}_2, m_2)$ is a space with endomorphism T_2 and if $T_1 \overset{\phi}{\to} T_2$ is a homomorphism then for any

$J_2 : X_2 \to \mathbb{R}$ $\Lambda_{T_2}(J_2) \subset \Lambda_{T_1}(J_2 \circ \phi)$. Hence $\Lambda_{T_2}(J_2) = \Lambda_{T_1}(J_2 \circ \phi)$ if ϕ is an isomorphism. The following proposition should be clear from these remarks and 7.

35. Proposition. If T_1, T_2 are regularly isomorphic processes with finite information cocycles then $\Lambda(T_1) = \Lambda(T_2)$.

36. Exercise. Let S, T be processes. If $(a, b) \in \Lambda(T)$ show that $(na, b) \in \Lambda(T^n)$. If $(na, b) \in \Lambda(T^n)$ show that $(na, nb) \in \Lambda(T)$. Also show that if $(a, c) \in \Lambda(S)$ and $(b, c) \in \Lambda(T)$ then $(a + b, c) \in \Lambda(S \times T)$. Can more be said?

We now prove some results toward the computation of $\Lambda(T)$.

37. Proposition. If $(X, \mathcal{B}, \mathcal{A}, m, T)$ is a process and if $F \circ T = fF$ where $|f| = 1$ and f is $\mathcal{A}$-measurable then F is $\mathcal{A}$-measurable.

Proof. First assume $F \in L^2(X, \mathcal{B}, m)$. By definition $\mathcal{A} \subset T\mathcal{A} \subset T^2\mathcal{A} \subset \ldots$ and $\bigcup_{n \geq 0} T^n\mathcal{A}$ generates $\mathcal{B}$ so $L^2(X, \mathcal{A}, m) \subset L^2(X, T\mathcal{A}, m) \subset L^2(X, T^2\mathcal{A}, m) \subset \ldots$ and $\bigcup_{n \geq 0} L^2(X, T^n\mathcal{A}, m)$ is dense in $L^2(X, \mathcal{B}, m)$. Denote by U the unitary operator defined by T, $Uf = f \circ T$ for $f \in L^2(X, \mathcal{B}, m)$. Put $V = L^2(X, T\mathcal{A}, m) \ominus L^2(X, \mathcal{A}, m)$ and note that for $n \in \mathbb{Z}$

$$U^n(L^2(X, \mathcal{A}, m)) = L^2(X, T^{-n}\mathcal{A}, m) \text{ and}$$

$$U^n(V) = L^2(X, T^{-n+1}\mathcal{A}, m) \ominus L^2(X, T^{-n}\mathcal{A}, m) .$$

We have $L^2(X, \mathcal{B}, m) = L^2(X, \mathcal{A}, m) \oplus \bigoplus_{n=0}^{\infty} U^{-n}V$ since the subspace on the right is closed and contains each $L^2(X, T^n\mathcal{A}, m)$, $n \geq 0$. Write $F = F_0 + \sum_{n=0}^{\infty} U^{-n}f_n$ where $F_0 \in L^2(X, \mathcal{A}, m)$, $f_n \in V$. By assumption,

$$F = \bar{f}(F \circ T) = \bar{f}(F_0 \circ T + f_0 \circ T) + \sum_{n=0}^{\infty} \bar{f}\, U^{-n} f_{n+1} .$$

Note that multiplication by $\bar{f} \in L^\infty(X, \mathcal{A}, m)$ maps $L^2(X, \mathcal{A}, m)$ to itself and each $U^{-n}V$ to itself and use uniqueness of the expression for f to deduce

$$F_0 = \bar{f}(F_0 \circ T + f_0 \circ T)$$

$$f_0 = \bar{f} f_1, \; f_1 \circ T^{-1} = \bar{f}(f_2 \circ T^{-1}), \; f_2 \circ T^{-1} = \bar{f}(f_3 \circ T^{-1}), \ldots$$

Thus $f_0, f_1, \ldots$ all have the same L^2 norm and, for the convergence of the series for F, must all be zero. Hence $F = F_0 \in L^2(X, \mathcal{A}, m)$.

For measurable $F : X \to \mathbb{C}$ truncate by putting $F_n = F\chi_{B_n}$ where $B_n = \{x : |F| \leq n\}$. Then $F_n \in L^\infty(X, \mathcal{B}, m) \subset L^2(X, \mathcal{B}, m)$. Since $|f| = 1$, $F_n \circ T = fF_n$. Thus each F_n is $\mathcal{A}$-measurable. But $F_n \to F$ (pointwise) and so F is $\mathcal{A}$-measurable. //

38. Proposition. Let T be an automorphism of $(X, \mathcal{B}, m)$ with generator α. If $F \circ T = fF$ where $|f| = 1$ and f is $\alpha \vee T^{-1}\alpha \vee \ldots \vee T^{-n}\alpha$ measurable, then F is measurable with respect to $(\bigvee_{i=0}^{\infty} T^{-i}\alpha) \cap (\bigvee_{i=-(n-1)}^{\infty} T^{i}\alpha)$.

Proof. f is $\bigvee_{i=0}^{\infty} T^{-i}\alpha$ measurable so F is $\bigvee_{i=0}^{\infty} T^{-i}\alpha$ measurable by 37. $F \circ T^{-1} = (\bar{f} \circ T^{-1})F$ and $\bar{f} \circ T^{-1}$ is measurable with respect to $\bigvee_{i=-(n-1)}^{1} T^{i}\alpha \subset \bigvee_{i=-(n-1)}^{\infty} T^{i}\alpha$, which is T^{-1} exhaustive. 37 shows that F is also $\bigvee_{i=-(n-1)}^{\infty} T^{i}\alpha$ measurable. //

39. Corollary. (i) If $(X, \mathcal{B}, \mathcal{A}, m, T)$ is a process and if $F \circ T = F + f$ where F, f are finite and $\mathcal{B}, \mathcal{A}$ measurable respectively, then F is $\mathcal{A}$ measurable.

(ii) If T is an automorphism of $(X, \mathcal{B}, m)$ with generator α and if $F \circ T = F + f$ where F, f are finite and f is $\alpha \vee T^{-1}\alpha \vee \ldots \vee T^{-n}\alpha$ measurable, then F is measurable with respect to $(\bigvee_{i=0}^{\infty} T^{-i}\alpha) \cap (\bigvee_{i=-(n-1)}^{\infty} T^{i}\alpha)$.

Proof. (i) The equation

$$\exp 2\pi it\, F \circ T = \exp 2\pi itf \,.\, \exp 2\pi\, itF$$

and 37 show that $\exp 2\pi itF$ is $\mathcal{A}$-measurable for every $t \in \mathbb{R}$. By approximation, $g \circ F$ is $\mathcal{A}$-measurable for any continuous $g : \mathbb{R} \to \mathbb{C}$ with compact support. Again by approximation, $F^{-1}(S) \in \mathcal{A}$ for each bounded interval S and the result follows.

(ii) is obtained from 38 in a similar way. //

40. Exercise. Prove the above corollary directly, under the additional assumption that F, f are square integrable.

41. Proposition. If $(X, \mathcal{B}, m, T)$ is a Markov chain with state partition α, then $(\bigvee_{i=0}^{\infty} T^{-i}\alpha) \cap (\bigvee_{i=-(n-1)}^{\infty} T^{i}\alpha) = \alpha \vee T^{-1}\alpha \vee \dots \vee T^{-(n-1)}\alpha$ for $n=1, 2, \dots$.

Proof. Since $\alpha \vee T^{-1}\alpha \vee \dots \vee T^{-(n-1)}\alpha$ may be considered as the state partition of an isomorphic Markov chain it suffices to prove the proposition for the case $n = 1$ i.e. $(\bigvee_{i=0}^{\infty} T^{-i}\alpha) \cap (\bigvee_{i=0}^{\infty} T^{i}\alpha) = \alpha$.

Let $f \in L^1$ be measurable with respect to the intersection σ-algebra. Put $f_n = E(f|\bigvee_{i=0}^{n} T^{i}\alpha)$. By the increasing Martingale theorem,

$$f_n \to E(f|\bigvee_{i=0}^{\infty} T^{i}\alpha) = f \text{ a.e. and in } L^1.$$

Since

$$\int |E(f|\bigvee_0^{\infty} T^{-i}\alpha) - E(f_n|\bigvee_0^{\infty} T^{-i}\alpha)|dm \le \int |f - f_n|dm,$$

$$E(f_n|\bigvee_0^{\infty} T^{-i}\alpha) \to E(f|\bigvee_0^{\infty} T^{-i}\alpha) = f \text{ in } L^1.$$

Put $f_n = \sum_C a(C)\chi_C$ where the sum is over all cylinders C of the form $[i_{-n}, i_{-n+1}, \dots, i_0]^{-n}$. Then

$$E(f_n|\bigvee_0^{k} T^{-i}\alpha) = \sum_D \chi_D(\int_D f_n\, dm)/m(D)$$

where the sum is over all cylinders D of the form $[j_0, \dots, j_k]$.

$$\text{I.e. } E(f_n|\bigvee_0^{k} T^{-i}\alpha) = \sum_D \chi_D \frac{1}{m(D)} \sum_C a(C)m(C \cap D)$$
$$= \sum_D \chi_D(\sum_C a(C)m(C \cap D)/m(D))$$

which is easily seen to be α measurable. Again by the increasing Martingale theorem $E(f_n|\bigvee_0^{k} T^{-i}\alpha) \to E(f_n|\bigvee_0^{\infty} T^{-i}\alpha)$ in L^1 as $k \to \infty$. Thus each $E(f_n|\bigvee_0^{\infty} T^{-i}\alpha)$ is α measurable, and so is their L^1 limit f. The proof is completed by considering $f = \chi_B$ for B in the intersection σ-algebra. //

42. Corollary. Let T be a Markov chain with state partition α. If $F \circ T = fF$ ($|f| = 1$) or $F \circ T = f + F$ where f is $\alpha \vee T^{-1}\alpha \vee \dots \vee T^{-n}\alpha$ measurable, then F is $\alpha \vee \dots \vee T^{-(n-1)}\alpha$ measurable.

In the same way one proves

43. Proposition. If T is Bernoulli with state partition α and if $F \circ T = fF$ where $|f| = 1$ and f is α measurable, then F is constant a.e. (i.e. $f = 1$).

We may now prove with little effort the following interesting

44. Proposition. For a Markov chain T the following are equivalent

(i) $\sigma^2(T) = 0$

(ii) I_T is cohomologous to a constant

(iii) T is of maximal type.

Proof. We easily see from 22 that (i) and (ii) are equivalent. We show (ii) and (iii) are equivalent. If T is defined by $P(i, j) = \frac{A(i, j) r(j)}{\beta r(i)}$ where A is a 0-1 matrix, β its maximum eigenvalue and r a corresponding eigenvector, then I_T is cohomologous to the function equal on $[i, j]$ to

$$-\log P(i, j) = \log \beta - \log r(j) + \log r(i)$$

i.e. I_T is cohomologous to the constant $\log \beta$. Conversely, if I_T is cohomologous to the constant $\log \beta$, $\beta > 0$, then

$$-\log P(i, j) = \log \beta + f \circ T - f$$

for some f. We see from 42 that f is then α measurable so that

$$P(i, j) = A(i, j) \frac{e^{-f(j)}}{\beta e^{-f(i)}}$$

where A is the 0-1 matrix compatible with P. Since P is stochastic, $\beta > 0$ is an eigenvalue of A with corresponding eigenvector $(e^{-f(1)}, \dots, e^{-f(k)})$. Hence, by Perron-Frobenius, β is the maximal eigenvalue of A and T is of maximal type. //

Consider the Markov chains T_1, T_2 and T_3 respectively defined by $\begin{pmatrix} p & q \\ p & q \end{pmatrix}$, $\begin{pmatrix} p & q \\ q & p \end{pmatrix}$ and $\begin{pmatrix} q & p \\ p & q \end{pmatrix}$ where $0 < p < 1$, $p \neq \frac{1}{2}$ and $p + q = 1$. These have the same entropy and, as we have seen in 18 and 20, the same information variance. It is also not hard to see that $P(tI_{T_1}) = P(tI_{T_2}) = P(tI_{T_3})$ for all $t \in \mathbb{R}$. We now show that T_1, T_2, T_3 are not regularly isomorphic by computing their Λ groups.

T_1 is the Bernoulli process given by (p, q) so that if $(a, b) \in \Lambda(T_1)$ then for some F with $|F| = 1$ we have $F \circ T = F \exp 2\pi i(a + bI_{T_1})$ and, by 43,

$$\exp 2\pi i(a + bI_{T_1}) = F \circ T/F = 1.$$

We conclude that $a + bI_{T_1}$ must take values in $\mathbb{Z}$ and that

$$\Lambda(T_1) = \{(a, b) \in \mathbb{R}^2 : a - b \log p,\ a - b \log q \in \mathbb{Z}\}.$$

If $(a, b) \in \Lambda(T_2)$ we have

$$F \circ T/F = \begin{cases} \exp 2\pi i(a - b \log p) & \text{on } [11] \cup [22] \\ \exp 2\pi i(a - b \log q) & \text{on } [12] \cup [21] \end{cases}$$

for some F with $|F| = 1$. 42 shows that F depends only on the zero coordinate so, $\exp 2\pi i(a - b \log p) = \frac{F(1)}{F(1)} = 1$ and $\exp 2\pi i(a - b \log q) = \frac{F(2)}{F(1)} = \frac{F(1)}{F(2)}$. Hence

$$\Lambda(T_2) = \{(a, b) \in \mathbb{R}^2 : a - b \log p,\ 2(a - b \log q) \in \mathbb{Z}\}.$$

Interchanging p and q, $\Lambda(T_3) = \{(a, b) \in \mathbb{R}^2 : 2(a - b \log p),\ a - b \log q \in \mathbb{Z}\}$. Thus $\Lambda(T_1)$, $\Lambda(T_2)$, $\Lambda(T_3)$ are pairwise distinct, and the processes T_1, T_2, T_3 are not regularly isomorphic.

6. QUASI-REGULAR ISOMORPHISMS AND BOUNDED CODES

In this section we relax the rather stringent requirement of regularity but insist that isomorphisms be sufficiently well-behaved (quasi-regular) to ensure that the cocycle-coboundary equation is retained at least in a weak form. Quasi-regularity is defined in terms of a metric on sub-σ-algebras, which we now describe. We fix for this section a Lebesgue space $(X, \mathcal{B}, m)$.

45. Definition. (i) Let $\alpha = (A_1, A_2, \ldots)$ be a (finite or) countable ordered partition of $(X, \mathcal{B}, m)$. If $\beta = (B_1, B_2, \ldots)$ is an ordered partition of the same cardinality, $d(\alpha, \beta) = \sum_n m(A_n \Delta B_n) = 2 - 2\sum_n m(A_n \cap B_n)$. If $\mathcal{C} \subset \mathcal{B}$ is a sub-σ-algebra $d(\alpha, \mathcal{C}) = \inf_{\gamma \subset \mathcal{C}} d(\alpha, \gamma)$ where the infimum is over all partitions of the same cardinality as α.

(ii) If $\mathcal{A}, \mathcal{C} \subset \mathcal{B}$ are sub-σ-algebras, $d(\mathcal{A}, \mathcal{C}) = \sup_{\alpha \subset \mathcal{A}} d(\alpha, \mathcal{C})$ where the supremum is over all countable partitions.

The relations listed in the following proposition are easy to verify or proofs can be found in [P6] or [P7].

46. Proposition. If $\mathcal{A}$, $\mathcal{C}$ etc. are sub-σ-algebras of $\mathcal{B}$, then

(i) $0 \leq d(\mathcal{A}, \mathcal{C}) \leq 2$,

(ii) $d(\mathcal{A}, \mathcal{C}) = 0$ iff $\mathcal{A} \subset \mathcal{C}$,

(iii) $d(\mathcal{A}_1, \mathcal{A}_2) \leq d(\mathcal{A}_1, \mathcal{A}_3) + d(\mathcal{A}_3, \mathcal{A}_2)$ so $d(\mathcal{A}_1, \mathcal{A}_2) \leq d(\mathcal{A}_3, \mathcal{A}_2)$ if $\mathcal{A}_1 \subset \mathcal{A}_3$ and $d(\mathcal{A}_1, \mathcal{A}_2) \leq d(\mathcal{A}_1, \mathcal{A}_3)$ if $\mathcal{A}_3 \subset \mathcal{A}_2$,

(iv) $d(\mathcal{A}_1 \vee \mathcal{A}_2, \mathcal{C}_1 \vee \mathcal{C}_2) \leq d(\mathcal{A}_1, \mathcal{C}_1) + d(\mathcal{A}_2, \mathcal{C}_2)$ so $d(\mathcal{A}_1 \vee \mathcal{A}_2, \mathcal{C}) = d(\mathcal{A}_1, \mathcal{C})$ when $\mathcal{A}_2 \subset \mathcal{C}$,

(v) $d(\mathcal{A}, \mathcal{C}) = d(T^{-1}\mathcal{A}, T^{-1}\mathcal{C})$ for a homomorphism (endomorphism) T into X,

(vi) $d(\mathcal{A}_n, \mathcal{C}) \uparrow d(\mathcal{A}, \mathcal{C})$ when $\mathcal{A}_n \uparrow \mathcal{A}$,

(vii) $d(\alpha, \mathcal{C}_n) \downarrow d(\alpha, \mathcal{C})$ when $\mathcal{C}_n \uparrow \mathcal{C}$ and α is a countable partition.

Despite 46(vi) and (vii), it is not true in general that $d(\mathcal{A}, \mathcal{C}_n) \to d(\mathcal{A}, \mathcal{C})$ when $\mathcal{C}_n \uparrow \mathcal{C}$. It is useful to note that $d(\mathcal{A}, \mathcal{C}) = \sup\{d(\alpha, \mathcal{C}) : \alpha \subset \mathcal{A}$ is a finite partition$\}$ since $(X, \mathcal{B}, m)$ is Lebesgue. 46(i)-(iii) show that d is nearly a metric - it is not symmetric. To get a metric we define $D(\mathcal{A}, \mathcal{C}) = \max\{d(\mathcal{A}, \mathcal{C}), d(\mathcal{C}, \mathcal{A})\}$ for sub-σ-algebras $\mathcal{A}, \mathcal{C} \subset \mathcal{B}$.

47. Theorem [P7]. Let $\mathcal{A}$, $\mathcal{C}$ be sub-σ-algebras. Then $d(\mathcal{A}, \mathcal{C}) < 2$ iff $I(\mathcal{A}|\mathcal{C})$ is finite on a set of positive measure.

Proof. Since $d(\mathcal{A} \vee \mathcal{C}, \mathcal{C}) = d(\mathcal{A}, \mathcal{C})$ and $I(\mathcal{A} \vee \mathcal{C}|\mathcal{C}) = I(\mathcal{A}|\mathcal{C})$, we may assume $\mathcal{C} \subset \mathcal{A}$. Suppose $d(\mathcal{A}, \mathcal{C}) < 2$. Pick $\varepsilon > 0$ small enough to have $d(\mathcal{A}, \mathcal{C}) < 2 - 2\varepsilon$. Let $\alpha = (A_1, A_2, \ldots) \subset \mathcal{A}$ be a countable partition. By the definition of d, there is a partition $\gamma = (C_1, C_2, \ldots) \subset \mathcal{C}$ such that

$$d(\alpha, \gamma) = \sum_n m(A_n \Delta C_n) < 2 - 2\varepsilon \quad \text{i.e.} \quad \sum_n m(A_n \cap C_n) > \varepsilon.$$

By Hölder's inequality,

$$\sum_n m(C_n \cap A_n|\mathcal{C}) = \sum_n \chi_{C_n} m(A_n|\mathcal{C}) \leq \left(\sum_n \chi_{C_n}\right)^{\frac{1}{2}} \left(\sum_n m(A_n|\mathcal{C})^2\right)^{\frac{1}{2}} = \left(\sum_n m(A_n|\mathcal{C})^2\right)^{\frac{1}{2}}$$

so that

$$\left(\int \sum_n m(A_n|\mathcal{C})^2 dm\right)^{\frac{1}{2}} \geq \int \left(\sum_n m(A_n|\mathcal{C})^2\right)^{\frac{1}{2}} dm$$

$$\geq \int \sum_n m(C_n \cap A_n | \mathcal{C})\, dm = \sum_n m(A_n \cap C_n) > \varepsilon .$$

Hence, for any partition $\alpha = (A_1, A_2, \dots) \subset \mathcal{A}$,

$$\int E(\alpha|\mathcal{C})\, dm > \varepsilon^2 \quad \text{where } E(\alpha|\mathcal{C}) = \sum_n \chi_{A_n} m(A_n|\mathcal{C}).$$

Now choose partitions $\alpha_k \uparrow \mathcal{A}$ and put $E(\mathcal{A}|\mathcal{C}) = \lim_{k\to\infty} E(\alpha_k|\mathcal{C})$ so that $I(\mathcal{A}|\mathcal{C}) = -\log E(\mathcal{A}|\mathcal{C})$. From above, $\int E(\mathcal{A}|\mathcal{C})\, dm > \varepsilon^2$. We conclude that $E(\mathcal{A}|\mathcal{C})$ is positive and $I(\mathcal{A}|\mathcal{C}) < \infty$, on a set of positive measure.

Conversely, suppose $I(\mathcal{A}|\mathcal{C}) < \infty$ on a set of positive measure. Then $E(\mathcal{A}|\mathcal{C})$ is positive on this set and we can find $\varepsilon > 0$ such that if $E = \{x : E(\mathcal{A}|\mathcal{C}) > \varepsilon\}$ then $m(E) > \varepsilon$. Let $\alpha \subset \mathcal{A}$ be a finite partition finer than (E, E^c) so that $E = A_1 \cup \dots \cup A_k$, a disjoint union of sets in α. Then $E(\alpha|\mathcal{C}) \geq E(\mathcal{A}|\mathcal{C})$ so, for each $1 \leq i \leq k$, $m(A_i|\mathcal{C}) > \varepsilon$ on A_i and, using the fact that $m(A_i|\mathcal{C})$ is $\mathcal{C}$ measurable, $m(A_i|\mathcal{C}) > \varepsilon$ on some $C_i \in \mathcal{C}$, $A_i \subset C_i$. Put $C_1' = C_1$, $C_2' = C_2 - C_1$, $C_3' = C_3 - (C_1 \cup C_2)$, ..., $C_k' = C_k - (C_1 \cup \dots \cup C_{k-1})$. C_i' are disjoint and $C_i \supset C_i'$ so

$$\sum_{i=1}^{k} m(C_i' \cap A_i) = \sum_{i=1}^{k} \int_{C_i'} m(A_i|\mathcal{C})\, dm$$

$$\geq \varepsilon \sum_{i=1}^{k} m(C_i') = \varepsilon\, m\Big(\bigcup_{i=1}^{k} C_i\Big) = \varepsilon\, m(E) > \varepsilon^2 .$$

Hence $d(\alpha, \mathcal{C}) < 2 - 2\varepsilon^2$ provided α refines (E, E^c) and it follows that $d(\mathcal{A}, \mathcal{C}) \leq 2 - 2\varepsilon^2$. //

48. Definition. Two sub-σ-algebras $\mathcal{A}$ and $\mathcal{C}$ are said to be quasi-regularly related if $D(\mathcal{A}, \mathcal{C}) < 2$ or, equivalently, if $I(\mathcal{A}|\mathcal{C})$ and $I(\mathcal{C}|\mathcal{A})$ are each finite on sets of positive measure. Two processes $(X_i, \mathcal{B}_i, \mathcal{A}_i, m_i, T_i)$ $(i = 1, 2)$ are said to be quasi-regularly isomorphic if there is an isomorphism $T_1 \xrightarrow{\phi} T_2$ such that $\mathcal{A}_1$ and $\phi^{-1}\mathcal{A}_2$ are quasi-regularly related.

This definition is justified by the following theorem and its corollaries.

49. Theorem [P. 7]. Let T be an ergodic endomorphism of $(X, \mathcal{B}, m)$ and let $\mathcal{A}$, $\mathcal{C}$ be quasi-regularly related sub-σ-algebras such that $T^{-1}\mathcal{A} \subset \mathcal{A}$, $T^{-1}\mathcal{C} \subset \mathcal{C}$. If $I(\mathcal{A}|T^{-1}\mathcal{A})$ is finite (a.e.) then $I(\mathcal{C}|T^{-1}\mathcal{C})$, $I(\mathcal{A}|\mathcal{C})$ and $I(\mathcal{C}|\mathcal{A})$ are finite (a.e.), and $I(\mathcal{A}|T^{-1}\mathcal{A})$ and $I(\mathcal{C}|T^{-1}\mathcal{C})$ are cohomologous.

<u>Proof.</u> We have $I(\mathcal{A} \vee \mathcal{C} | T^{-1}\mathcal{A}) = I(\mathcal{A} | T^{-1}\mathcal{A}) + I(\mathcal{C} | \mathcal{A})$,

$$I(\mathcal{A} \vee \mathcal{C} | T^{-1}\mathcal{A}) = I(\mathcal{A} \vee \mathcal{C} | T^{-1}\mathcal{A} \vee T^{-1}\mathcal{C}) + I(\mathcal{C} | \mathcal{A}) \circ T.$$

Put $E = \{x : I(\mathcal{C} | \mathcal{A}) < \infty\}$ and $F = \{x : I(\mathcal{A} \vee \mathcal{C} | T^{-1}\mathcal{A} \vee T^{-1}\mathcal{C}) < \infty\}$. By 47, $m(E) > 0$. Since $I(\mathcal{A} | T^{-1}\mathcal{A})$ is finite by assumption, the above equalities show that $m(F) \geq m(E) > 0$. Also $F \cap T^{-1}E = E$ so that $T^{-1}E \supset E$ and, as T is (measure-preserving and) ergodic, $m(E) = 1$. It follows that $m(F) = 1$. Thus $I(\mathcal{C} | \mathcal{A})$ and $I(\mathcal{A} \vee \mathcal{C} | T^{-1}\mathcal{A} \vee T^{-1}\mathcal{C})$ are finite (a.e.). The equations

$$I(\mathcal{A} \vee \mathcal{C} | T^{-1}\mathcal{C}) = I(\mathcal{C} | T^{-1}\mathcal{C}) + I(\mathcal{A} | \mathcal{C})$$
$$= I(\mathcal{A} \vee \mathcal{C} | T^{-1}\mathcal{A} \vee T^{-1}\mathcal{C}) + I(\mathcal{A} | \mathcal{C}) \circ T$$

show by a similar argument that $I(\mathcal{A} | \mathcal{C})$ and $I(\mathcal{C} | T^{-1}\mathcal{C})$ are finite (a.e.). That $I(\mathcal{A} | T^{-1}\mathcal{A})$ and $I(\mathcal{C} | T^{-1}\mathcal{C})$ are cohomologous can be seen from the four equations above. //

50. <u>Corollary.</u> <u>If</u> T <u>is an ergodic endomorphism of</u> $(X, \mathcal{B}, m)$, <u>then the quasi-regularity relation is an equivalence relation between sub-</u>σ<u>-algebras</u> $\mathcal{A}$ <u>with</u> $T^{-1}\mathcal{A} \subset \mathcal{A}$ <u>and</u> $I(\mathcal{A} | T^{-1}\mathcal{A}) < \infty$ <u>a.e.</u>

<u>Proof.</u> Reflexivity and symmetry are clear. To show that the relation is transitive suppose $D(\mathcal{A}_1, \mathcal{A}_2) < 2$ and $D(\mathcal{A}_2, \mathcal{A}_3) < 2$ where $\mathcal{A}_i \subset \mathcal{B}$ satisfy $T^{-1}\mathcal{A}_i \subset \mathcal{A}_i$, $I(\mathcal{A}_i | T^{-1}\mathcal{A}_i) < \infty$ a.e. for $i = 1, 2, 3$. Note that

$$I(\mathcal{A}_1 | \mathcal{A}_3) \leq I(\mathcal{A}_1 \vee \mathcal{A}_2 | \mathcal{A}_3) = I(\mathcal{A}_1 | \mathcal{A}_2 \vee \mathcal{A}_3) + I(\mathcal{A}_2 | \mathcal{A}_3)$$

and $I(\mathcal{A}_2 | \mathcal{A}_3) < \infty$ a.e. by 49. Since $d(\mathcal{A}_1, \mathcal{A}_2 \vee \mathcal{A}_3) \leq d(\mathcal{A}_1, \mathcal{A}_2) < 2$, $I(\mathcal{A}_1 | \mathcal{A}_2 \vee \mathcal{A}_3)$ and $I(\mathcal{A}_1 | \mathcal{A}_3)$ are finite on a set of positive measure. Hence $d(\mathcal{A}_1, \mathcal{A}_3) < 2$. Similarly $d(\mathcal{A}_3, \mathcal{A}_1) < 2$. //

51. <u>Corollary.</u> <u>If two ergodic processes</u> $(X_i, \mathcal{B}_i, \mathcal{A}_i, m_i, T_i)$ $(i = 1, 2)$ <u>are quasi-regularly isomorphic by</u> $T_1 \xrightarrow{\psi} T_2$ <u>then</u> $I_{T_1} < \infty$ <u>a.e. iff</u> $I_{T_2} < \infty$ <u>a.e. and, in this case,</u> I_{T_1}, $I_{T_2} \circ \phi$ <u>are cohomologous.</u>

52. <u>Corollary.</u> <u>If two ergodic processes with finite information cocycles are quasi-regularly isomorphic then one of the cocycles is cohomologous to a constant iff the other is. Consequently, if the processes are Markov, then one of them is of</u>

maximal type iff the other is.

Quasi-regularity is related to, and was suggested by, Bowen's notion of bounded coding [B. 2]:

53. Definition. When T is an automorphism of $(X, \mathcal{B}, m)$ and α, β are finite partitions, β boundedly codes α (with respect to T) if for every $\varepsilon > 0$ there exists $k \in \mathbb{N}$ such that for all $n \in \mathbb{N}$

$$d\left(\bigvee_{i=0}^{n} T^{-i}\alpha, \bigvee_{i=-k}^{n+k} T^{-i}\beta\right) < \varepsilon .$$

When T_i $(i = 1, 2)$ are two finite state processes with state partitions α_i, then T_2 boundedly codes T_1 if there is an isomorphism $T_1 \xrightarrow{\phi} T_2$ such that $\phi^{-1}\alpha_2$ boundedly codes α_1 with respect to T_1.

54. Proposition. If the finite state process T_2 boundedly codes the finite state process T_1 through the isomorphism $T_1 \xrightarrow{\phi} T_2$ then $d(\mathcal{A}_1, \varphi^{-1}\mathcal{A}_2) < 2$, where $\mathcal{A}_i = \bigvee_{j=0}^{\infty} T_i^{-j}\alpha_i$, α_i state partitions.

Proof. It is easy to see that for $0 < \varepsilon < 2$, there exists $k \in \mathbb{N}$ such that

$$d\left(\bigvee_{i=0}^{n} T_1^{-i}\alpha_1, T_1^{k}\phi^{-1}\mathcal{A}_2\right) < \varepsilon$$

for $n = 1, 2, \ldots$. Therefore, by 46(vi), $d(\mathcal{A}_1, T_1^{k}\phi^{-1}\mathcal{A}_2) \leq \varepsilon < 2$. Hence $I(\mathcal{A}_1 | T_1^{k}\phi^{-1}\mathcal{A}_2)$ is finite on a set of positive measure. It follows that $I(\mathcal{A}_1 | \phi^{-1}\mathcal{A}_2)$ is also finite on a set of positive measure, since

$$\begin{aligned} I(\mathcal{A}_1 | \phi^{-1}\mathcal{A}_2) &\leq I(\mathcal{A}_1 \vee T_1^{k}\phi^{-1}\mathcal{A}_2 | \phi^{-1}\mathcal{A}_2) \\ &= I(\mathcal{A}_1 | T_1^{k}\phi^{-1}\mathcal{A}_2) + I(\varphi^{-1}\mathcal{A}_2 | T_1^{-k}\phi^{-1}\mathcal{A}_2) \circ T_1^{-k} \end{aligned}$$

and the last term has integral

$$H(\alpha_2 \vee \ldots \vee T_2^{-(k-1)}\alpha_2 | T_2^{-k}\mathcal{A}_2) \leq kH(\alpha_2) < \infty . \;//$$

55. Corollary. If the finite state processes T_1, T_2 simultaneously boundedly code each other (i.e. if there exists an isomorphism $T_1 \xrightarrow{\psi} T_2$ such that α_1 and $\psi^{-1}\alpha_2$ boundedly code each other), then T_1, T_2 are quasi-regularly isomorphic.

In view of the above results the group invariant of the preceding section can be used to distinguish ergodic processes from the point of view of quasi-regular isomorphism (and finite state processes from the point of view of bounded coding). Unlike theorem 7, 51 does not indicate when the information cocycles are L^q-cohomologous $(1 \leq q \leq \infty)$. The pressure invariants of Section 4 cannot be used unless we have some means of guaranteeing that the information cocycles are cohomologous by an L^∞-coboundary (in the sense of definition 5). Information variance, however, can be used even when the functions occurring in 51 are not known to be in L^2. This, as we shall see in the next section, is because information variance presents itself in various central limit theorems.

7. CENTRAL LIMITING DISTRIBUTIONS AS INVARIANTS

In this section we show how a central limiting distribution (if it exists) can be used as an invariant of the relations

$$T_1 \xrightarrow{\phi} T_2, \quad I^0_{T_1} = I^0_{T_2} \circ \phi + g \circ T_1 - g \qquad (\star)$$

for processes T_1, T_2. We assume that these relations hold and that the functions which appear in $(\star)$ are finite a. e.

56. Proposition. Suppose the relations $(\star)$ hold and for a sequence $\theta(n) \to \infty$ consider

$$F_{T_i}(t) = \lim_{n\to\infty} m\{x : \frac{1}{\theta(n)}(I^0_{T_i} + \ldots + I^0_{T_i} \circ T_i^{n-1})(x) < t\}.$$

Then $F_{T_1}(t)$ exists iff $F_{T_2}(t)$ exists and, in this case, $F_{T_1}(t) = F_{T_2}(t)$.

Proof. This is immediate from the easily verified fact that $(g \circ T_1^n - g)/\theta(n) \to 0$ in measure. //

57. Corollary. If T_1, T_2 are quasi-regularly isomorphic Markov chains then $\sigma^2(T_1) = \sigma^2(T_2)$.

Proof. Since $I^0_{T_1}$, $I^0_{T_2}$ are functions of the zero and first coordinates, $F_{T_1}(t)$, $F_{T_2}(t)$ exist for $\theta(n) = n^{\frac{1}{2}}$ as long as $\sigma^2(T_1)$, $\sigma^2(T_2)$ are not zero. In this case,

$$F_{T_i}(t) = (2\pi\sigma^2(T_i))^{-\frac{1}{2}} \int_{-\infty}^{t} \exp(-u^2/2\sigma^2(T_i))\, du$$

Hence $\sigma^2(T_1) = \sigma^2(T_2)$ when they are not zero. If, say, $\sigma^2(T_1) = 0$, then by 44 I_{T_1} is cohomologous to a constant and therefore I_{T_2} is cohomologous to a constant and $\sigma^2(T_2) = 0$. The proof is now complete. //

Bowen proved in [B. 2] that sufficiently small smooth partitions for C^2 Anosov diffeomorphisms T preserving a probability equivalent to Lebesgue measure boundedly code each other. Therefore, if α is such a partition, $\bigvee_{i=0}^{\infty} T^{-i}\alpha$ is canonical in the sense that any other such partition β gives rise to a σ-algebra $\bigvee_{i=0}^{\infty} T^{-i}\beta$ which is intimately (quasi-regularly) related to it. This statement can be put into geometrical language; we simply stress that in this situation $I(\alpha | \bigvee_{i=1}^{\infty} T^{-i}\alpha)$ and $I(\beta | \bigvee_{i=1}^{\infty} T^{-i}\beta)$ are cohomologous. For the terms left undefined in this paragraph, we refer the reader to [B. 1] and [B. 2].

58. <u>Remarks and Problems.</u> (i) The idea of using limiting distributions as invariants is due, in a slightly different form, to Bowen [B. 2] although implicitly it occurs in [F. P.].

(ii) Is it possible in contexts more general than the above to canonically associate information cocycles to diffeomorphisms of compact manifolds preserving smooth probabilities?

(iii) The main open problem posed by this chapter is to find a complete set of invariants for regular and quasi-regular classifications of processes.

(iv) The problem of classification of endomorphisms is not dealt with in these notes. The strict isomorphism problem, as opposed to the shift equivalence problem, is considered in [V. 1], [V. 2], [K. M. T.] and [P. W.]. Important work on the representation of endomorphisms as factors of Bernoulli endomorphisms appears in [R']. The information function $I_S = I(\mathcal{B} | S^{-1}\mathcal{B})$ of an <u>endomorphism</u> S of $(X, \mathcal{B}, m)$ is clearly an isomorphism invariant which can be used directly, without the complications of an additional coboundary. Indeed, many endomorphisms are completely characterised by the multivariate distributions of I_S, $I_S \circ S$, $I_S \circ S^2$, ... (see [P. P. W.]).

CHAPTER III: FINITARY ISOMORPHISMS

1. THE MARKER METHOD ([K. S. 1], [K. S. 2])

1. Definition. Let $(X_i, \mathcal{B}_i, m_i, T_i)$ $(i = 1, 2)$ be countable state processes with state partitions α and β. A homomorphism $T_1 \xrightarrow{\varphi} T_2$ is called a finitary homomorphism (or a finitary code) if each $\varphi^{-1}B$, $B \in \beta$, can be expressed as a countable union of α-cylinders. When φ is an isomorphism, φ is called a finitary isomorphism if both φ and φ^{-1} are finitary codes.

It is easy to see from the definition that a homomorphism $T_1 \xrightarrow{\varphi} T_2$ of countable state processes is finitary iff for (almost all) $x = (x_n) \in X_1$ we can find integers $k, l \geq 0$ such that φ has constant zero coordinate on the cylinder $[x_{-l}, \ldots, x_0, \ldots, x_k]^{-l}$. The integers k, l will, in general, depend on the point x.

In the papers [K. S. 1] and [K. S. 2], Keane and Smorodinsky introduced a method of constructing finitary codes. This method is loosely termed the 'Marker Method'. It has since been used in a number of other papers (see for instance [A. J. R.], [J], [K. S. 3], [P']). In this section we illustrate the method in proving

2. Theorem [K. S. 2]. Two finite state Bernoulli processes with at least three states and the same entropy are finitarily isomorphic.

The section is based on [K. S. 2]. In order to avoid digressing half way through the proof of 2, we state first

3. Definition. Let U and V be finite sets with probabilities ρ and σ (on their power sets). A society from U to V is a map S from U to the power set of V with the property that for any set $B \subset U$, $\rho(B) \leq \sigma(S(B))$ where $S(B) = \bigcup_{b \in B} S(b)$. If R and S are societies from U to V, write $R < S$ if $R(b) \subset S(b)$ for all $b \in U$. For a society S from U to V, define the dual society $S^\star$ from V to U by letting $b \in S^\star(g)$ iff $g \in S(b)$. If S_i are societies from U_i to V_i $(1 \leq i \leq j)$, define the product society $S_1 \times \ldots \times S_j$ from

$U_1 \times \ldots \times U_j$ to $V_1 \times \ldots \times V_j$ by setting

$$(S_1 \times \ldots \times S_j)(b_1, \ldots, b_j) = S_1(b_1) \times \ldots \times S_j(b_j) .$$

4. Proposition [K. S. 1]. (i) $S^\star$ is a society.

(ii) $S_1 \times \ldots \times S_j$ is a society. Also, $(S_1 \times \ldots \times S_j)^\star = S_1^\star \times \ldots \times S_j^\star$.

(iii) For any society S from U to V there is a society $R < S$ such that

$$\text{card}\{g \in V : \exists b_1, b_2 \in U, b_1 \neq b_2, \text{ with } g \in R(b_1) \cap R(b_2)\} < \text{card } U.$$

Proof. (i) Let $G \subset V$, then $b \notin S^\star(G)$ iff $G \cap S(b) = \emptyset$, so that $S^\star(G)^C = \{b \in U : S(b) \subset G^C\}$. Hence

$$\rho(S^\star(G)^C) = 1 - \rho(S^\star(G)) = \rho\{b : S(b) \subset G^C\} \leq \sigma(G^C) = 1 - \sigma(G)$$

and it follows that $\sigma(G) \leq \rho(S^\star(G))$.

The second part of (ii) is straightforward. The first part of (ii) and (iii) may be found in [K. S. 1], and their proofs read independently of the rest of the paper. //

The following lemmas help reduce the proof of 2.

5. Lemma. (i) Fix $0 < x_0 < 1$. The entropy of the probability vector (x_0, y_1, y_2) increases as it becomes more uniform (i.e. (x_0, y_1, y_2) has greater entropy than (x_0, z_1, z_2) when $|\frac{1-x_0}{2} - y_1| < |\frac{1-x_0}{2} - z_1|$).

(ii) If $M = \sum_{i=1}^{k} m_i$, $m_i > 0$ and $M < 1$, then $-\sum_{i=1}^{k} m_i(\log m_i) \geq -M(\log M)$.

(iii) Suppose $r_0, r_1 > 0$, $r_0 + r_1 < 1$ and that the entropy of the probability vector $(r_0, r_1, 1-r_0-r_1)$ is strictly less than some $h > 0$. Then $1 - r_0 - r_1$ can be split to obtain a probability vector $(r_0, r_1, \ldots, r_k)$ with entropy h.

Proof. (i) follows from the fact that the function

$$-x \log x - (1 - x_0 - x) \log(1 - x_0 - x), \quad 0 < x < 1 - x_0,$$

is concave, has its maximum at the point $x = \frac{1-x_0}{2}$ and is symmetric about this point. (ii) follows from the fact that $-\log x$ $(x > 0)$ is convex and decreasing. For (iii) note that the vector $(r_0, r_1, \frac{1-r_0-r_1}{n}, \ldots, \frac{1-r_0-r_1}{n})$ with $n + 2$ entries has entropy tending to infinity with n, choose n large enough for its entropy to exceed h and use a connectedness-continuity argument. //

6. Lemma [K. S. 2]. Let $p = (p_0, \ldots, p_{a-1})$ and $q = (q_0, \ldots, q_{b-1})$ be probability vectors with the same entropy h and with $a, b \geq 3$. Then we can find a probability vector $r = (r_0, \ldots, r_{c-1})$ such that $h(r) = h$, $c \geq 3$ and $r_0 \in \{p_0, \ldots, p_{a-1}\}$, $r_1 \in \{q_0, \ldots, q_{b-1}\}$.

Proof. Assume without loss of generality that $p_0 \geq p_1 \geq \ldots \geq p_{a-1}$, $q_0 \geq q_1 \geq \ldots \geq q_{b-1}$ and $p_{a-1} \geq q_{b-1}$. Take $r_0 = p_0$, $r_1 = q_{b-1}$. Then $(r_0, r_1, 1-r_0-r_1)$ is less uniform than $(p_0, p_{a-1}, 1-p_0-p_{a-1})$ and, by 5(i) and (ii),

$$h(r_0, r_1, 1-r_0-r_1) \leq h(p_0, p_{a-1}, 1-p_0-p_{a-1}) \leq h(p) = h.$$

Now use 5(iii). //

Lemma 6 shows that, to prove 2, it is sufficient to construct a finitary isomorphism between Bernoulli processes defined by $p = (p_0, \ldots, p_{a-1})$, $q = (q_0, \ldots, q_{b-1})$ where $p_0 = q_0$, $a, b \geq 3$ and $h(p) = h(q)$. The state 0 will be used as a 'marker' - the finitary isomorphism ϕ will have the property that, for $x = (x_n)$, $x_n = 0$ iff $(\phi x)_n = 0$. We now set up the machinery for the construction of the finitary isomorphism.

Let $p = (p_0, \ldots, p_{a-1})$ be a probability vector, and let $(X, \mathcal{B}, m, T)$ be the Bernoulli process defined by it. Fix a sequence of integers, $0 < N_1 < N_2 < \ldots$. This sequence will be specified later.

7. Definition [K. S. 2]. A skeleton s is a positive integer r together with a configuration of finite sequences of zeros and 'holes'

$$0^{n_0} \underset{l_1}{\text{---}} 0^{n_1} \underset{l_2}{\text{---}} \cdots \underset{l_m}{\text{---}} 0^{n_m}$$

with $n_i \geq 1$ $(0 \leq i \leq m)$, $l_i \geq 1$ $(1 \leq i \leq m)$ and $\max_{1 \leq i \leq m-1} \{n_i\} < N_r \leq \min\{n_0, n_m\}$. Each n_i specifies the length of a block of zeros and each l_i specifies the length of a sequence of holes. $r = r(s)$ is called the rank of the skeleton s. The length of s is $l(s) = l_1 + \ldots + l_m$. A skeleton s' is a subskeleton of s if it has configuration

$$0^{n_t} \underset{l_{t+1}}{\text{---}} 0^{n_{t+1}} \underset{l_{t+2}}{\text{---}} \cdots \underset{l_{t'}}{\text{---}} 0^{n_{t'}}$$

for suitable $0 \leq t < t' \leq m$, and if $r(s') \leq r(s)$. Two subskeletons of s are said to be disjoint if their corresponding indices $\{t+1, \ldots, t'\}$ are disjoint (i.e. if their holes are). Subskeletons $s_1, \ldots, s_j$ form a decomposition of s if they are pairwise disjoint and the union of their indices is $\{1, \ldots, m\}$; write $s = s_1 \times \ldots \times s_j$ after ordering according to these indices.

Note that a configuration of zeros and holes may be given more than one rank, we distinguish between these as skeletons (with different ranks).

8. Proposition. Each skeleton s of rank $r > 1$ admits a unique decomposition into subskeletons of rank $(r - 1)$, called the rank decomposition of s.

Proof. If s is as in 7, the number of subskeletons in the rank decomposition is given by the number of i with $1 \leq i \leq m$, $n_i \geq N_{r-1}$ (the number of 'long' blocks of zeros). Start at 0^{n_0} and scan to the right until the first 'long' block is reached, take this as the first subskeleton. Now scan until the next 'long' block to obtain the next subskeleton. Continue until 0^{n_m} is reached. Uniqueness is clear, since no 'long' block (with $n_i \geq N_{r-1}$) can be internal to a subskeleton of rank $(r - 1)$. //

We say that a skeleton s appears in $x \in X$ if the holes of s may be filled with elements of $\{1, 2, \ldots, a-1\}$ to obtain a finite segment of x which is neither preceded nor followed by a zero.

9. Proposition. (i) For almost all $x = (x_n) \in X$, either $x_0 = 0$ or there exists for each $r \geq 1$ a unique skeleton $s_r(x)$ of rank r which appears in x around the zero coordinate.

(ii) Given any sequence $L_1, L_2, \ldots$ of lengths, we can choose $0 < N_1 < N_2 < \ldots$ such that for almost all $x \in X$ with $x_0 \neq 0$ we have $l(s_r(x)) \geq L_r$ for large enough r.

Proof. (i) Suppose $x_0 \neq 0$. To say that we can find a skeleton of rank r around the zero coordinate means that we can find blocks of zeros of length not less than N_r to the left and right of x_0. This is assured a.e. by the independence of the process. Uniqueness is clear since the skeleton must end at the first block of zeros whose length is greater than or equal to N_r.

(ii) In fact we show that $0 < N_1 < N_2 < \ldots$ can be chosen so that for almost all $x = (x_n) \in X$ with $x_0 \neq 0$, eventually $l(s_r(x))^- \geq L_r$ where $l(s_r(x))^-$ is the 'past' length of $s_r(x)$ i.e. the number of holes to the right of the zero coordinate place. Take $\varepsilon > 0$, small. For each r choose $N_r > \frac{(p_0 + \varepsilon) L_r}{1 - p_0 - \varepsilon}$, also ensuring $0 < N_1 < N_2 < \ldots$. By the ergodic theorem, for almost all $x \in X$ we have

$$\frac{\text{no. of occurrences of 0 in } (x_1, \ldots, x_n)}{n} \to p_0 \qquad (\star)$$

Suppose $x_0 \neq 0$ and $l(s_r(x))^- < L_r$ for some $r \geq 1$. Then

$$\frac{N_r + N}{l(s_r(x))^- + N_r + N} \geq \frac{N_r}{L_r + N_r} > (p_0 + \varepsilon) ,$$

where N is the number of zeros in $s_r(x)$ between the zero coordinate place and the right end block. Thus, if $l(s_r(x))^- < L_r$ for infinitely many r then x violates $(\star)$. //

From $p = (p_0, \ldots, p_{a-1})$ we obtain a new probability vector $\bar{p} = (\frac{p_1}{1-p_0}, \frac{p_2}{1-p_0}, \ldots, \frac{p_{a-1}}{1-p_0})$. Write $g = g(p)$ for the entropy of $\bar{p}$ and denote by μ the measure given to the set $\{1, \ldots, a-1\}$ by $\bar{p}$. Set

$$\eta = \min_{1 \leq i \leq a-1} \{\frac{p_i}{1-p_0}\}, \quad \theta = \max_{1 \leq i \leq a-1} \{\frac{p_i}{1-p_0}\}.$$

10. Definition. Given a skeleton s of length l, let $\mathfrak{F}(s) = \{1, \ldots, a-1\}^l$ and give this set the product of the measures μ on $\{1, \ldots, a-1\}$. $\mathfrak{F}(s)$ is called the filler set of s and the product measure, also denoted μ, is called the filler measure. An element of $\mathfrak{F}(s)$ is a filler of s. $g = g(p)$ is called the filler entropy of $(X, \mathcal{B}, m, T)$.

It is important to note that for a skeleton s of rank r and length l and $F = (f_1, \ldots, f_l) \in \mathfrak{F}(s)$, $\mu(F)$ is a conditional measure: $\mu(F)$ is the conditional measure that F is the filler of s determined by $x \in X$, given that $s_r(x) = s_r$.

Let $\{\varepsilon_r\}$ be an arbitrary sequence satisfying $1 > \varepsilon_1 > \varepsilon_2 > \ldots > 0$ and decreasing to zero. We use it to define, for each skeleton s, an equivalence relation on $\mathfrak{F}(s)$. The definition is by induction on $r(s)$:

Let s be a skeleton of rank 1 and length l. For $F = (f_1, \ldots, f_l) \in \mathfrak{F}(s)$ set

$$J(F) = \{1 \leq i \leq l : \mu(f_1)\mu(f_2) \ldots \mu(f_{i-1}) \geq \frac{1}{\eta} e^{-g(1-\varepsilon_1)l}\}.$$

For $i = 1$ the condition becomes $\eta \geq e^{-g(1-\varepsilon_1)l}$, and we see that either $J(F) = \emptyset$

for all $F \in \mathfrak{F}(s)$ or $1 \in J(F)$ for all $F \in \mathfrak{F}(s)$. For $F = (f_1, \ldots, f_l)$, $F' = (f_1', \ldots, f_l') \in \mathfrak{F}(s)$ put $F \sim F'$ iff $f_i = f_i'$ for all $i \in J(F)$. It can be checked that $F \sim F'$ implies $J(F) = J(F')$, so that $\sim$ is a well defined equivalence relation.

Suppose the $J(F)$ and the equivalence relations have been defined for all skeletons of rank less than r. Let s be a skeleton with rank r and length l. Consider the rank decomposition of s, $s = s_1 \times \ldots \times s_j$, given by 8 and identify $\mathfrak{F}(s) = \mathfrak{F}(s_1) \times \ldots \times \mathfrak{F}(s_j)$. For $F = (f_1, \ldots, f_l) = (F_1, \ldots, F_j)$ where $F_1, \ldots, F_j$ are the subfillers for $s_1, \ldots, s_j$, put $\tilde{J}(F) = \bigcup_{i=1}^{j} J(F_i)$. Letting $k_1, \ldots, k_v$ be the increasing enumeration of the complement of $\tilde{J}(F)$ in $\{1, \ldots, l\}$, set

$$J(F) = \tilde{J}(F) \cup \{k_i : 1 \le i \le v,\ (\prod_{k \in \tilde{J}(F)} \mu(f_k))\mu(f_{k_1}) \ldots \mu(f_{k_{i-1}}) \ge \frac{1}{\eta} e^{-g(1-\varepsilon_r)l}\}.$$

For $F = (f_1, \ldots, f_l)$, $F' = (f_1', \ldots, f_l') \in \mathfrak{F}(s)$ define $F \sim F'$ iff $f_k \sim f_k'$ for all $k \in J(F)$. It can be checked by induction that for all s and $F, F' \in \mathfrak{F}(s)$ $F \sim F'$ implies $J(F) = J(F')$ and thus that the equivalence relations are well defined. For s, let $\tilde{\mathfrak{F}}(s) = \mathfrak{F}(s)/\sim$ and denote the class of $F \in$ (s) by $\tilde{F}$. If $r(s) > 1$ and $s = s_1 \times \ldots \times s_j$ is the rank decomposition of s, write $\overline{\mathfrak{F}}(s) = \prod_{i=1}^{j} \tilde{\mathfrak{F}}(s_i)$. Denote the class of F in $\overline{\mathfrak{F}}(s)$ by $\overline{F}$. It is easy to see that

$$\overline{F} = \{F' = (f_1', \ldots, f_l') \in \mathfrak{F}(s) : f_i' = f_i \text{ for all } i \in \tilde{J}(F)\}.$$

It follows that each element of $\overline{\mathfrak{F}}(s)$ is a disjoint union of elements of $\tilde{\mathfrak{F}}(s)$. We give $\tilde{\mathfrak{F}}(s)$ and $\overline{\mathfrak{F}}(s)$ the (quotient) measures obtained by considering each equivalence class as a subset of $\mathfrak{F}(s)$.

11. Lemma. For a skeleton s of rank r and length l,

(i) $\mu(\tilde{F}) \ge e^{-g(1-\varepsilon_r)l}$ for all $F \in \mathfrak{F}(s)$.

(ii) $\text{card}\, \overline{\mathfrak{F}}(s) \le e^{g(1-\varepsilon_{r-1})l}$ whenever $r > 1$.

Proof. (i) We use induction. Let s be of rank 1 and length l, and let $F = (f_1, \ldots, f_l) \in \mathfrak{F}(s)$. If $J(F) = \emptyset$ then $\tilde{F} = \mathfrak{F}(s)$ and the inequality is clear. If $J(F) = \{1, \ldots, i\}$ then

$$\mu(\tilde{F}) = \mu(f_1) \ldots \mu(f_i) \ge \frac{\mu(f_i)}{\eta} e^{-g(1-\varepsilon_1)l} \ge e^{-g(1-\varepsilon_1)l},$$

by the definition of $J(F)$. Assume (i) for ranks less than r. Let s have rank r, length l and rank decomposition $s = s_1 \times \ldots \times s_j$. Let $F = (f_1, \ldots, f_l) \in \mathcal{F}(s)$. If $J(F) = \tilde{J}(F)$ then

$$\mu(\tilde{F}) \geq \prod_{i=1}^{j} e^{-g(1-\varepsilon_{r-1}) l_i} \quad \text{(where } l_i \text{ is the length of } s_i\text{)}$$

$$= e^{-g(1-\varepsilon_{r-1}) l} \geq e^{-g(1-\varepsilon_r) l} .$$

If $J(F) = \tilde{J}(F) \cup \{k_1, \ldots, k_i\}$ then

$$\mu(\tilde{F}) = \Big(\prod_{k \in \tilde{J}(F)} \mu(f_k)\Big)\, \mu(f_{k_1}) \ldots \mu(f_{k_{i-1}})\, \mu(f_{k_i})$$

$$\geq \frac{\mu(f_{k_i})}{\eta} e^{-g(1-\varepsilon_r) l} \geq e^{-g(1-\varepsilon_r) l} .$$

(ii) Let s have rank $r > 1$ and length l. Let $s_1 \times \ldots \times s_j$ be its rank decomposition, where s_i has length l_i. From (i), for each $1 \leq i \leq j$,

$$\text{card}\, \tilde{\mathcal{F}}(s_i) \leq e^{g(1-\varepsilon_{r-1}) l_i} \quad \text{so that}$$

$$\text{card}\, \overline{\mathcal{F}}(s) = \prod_{i=1}^{j} \text{card}\, \tilde{\mathcal{F}}(s_i) \leq e^{g(1-\varepsilon_{r-1}) l} . \;//$$

12. <u>Lemma.</u> (i) <u>For any</u> $\delta > 0$ <u>and</u> $r \geq 1$, <u>there exists</u> $l_0 = l_0(\delta, r)$ <u>such that if</u> s <u>is a skeleton of rank</u> r <u>and length</u> $l \geq l_0$, <u>then for</u> $F \in \mathcal{F}(s)$ <u>outside a set of measure at most</u> δ,

$$\mu(\tilde{F}) \leq \frac{1}{\eta} e^{-g(1-\varepsilon_r) l} .$$

(ii) <u>For any</u> $\delta > 0$ <u>and</u> $r \geq 1$, <u>there exists</u> $l_1 = l_1(\delta, r)$ <u>such that if</u> s <u>is a skeleton of rank</u> r <u>and length</u> $l \geq l_1$, <u>then for</u> $F \in \mathcal{F}(s)$ <u>outside a set of measure at most</u> δ <u>we have,</u>

$$\frac{\text{card}\, J(F)}{l} \geq 1 - \frac{2g}{|\log \theta|} \varepsilon_r - \frac{\log \eta}{l \log \theta} .$$

<u>Proof.</u> (i) Let $F = (f_1, \ldots, f_l) \in \mathcal{F}(s)$. If $J(F) \neq \{1, \ldots, l\}$ then, by definition, $\mu(\tilde{F}) = \Big(\prod_{i \in J(F)} \mu(f_i)\Big) < \frac{1}{\eta} e^{-g(1-\varepsilon_r) l}$. If $J(F) = \{1, \ldots, l\}$, then $\Big(\prod_{i \in J} \mu(f_i)\Big) \geq \frac{1}{\eta} e^{-g(1-\varepsilon_r) l}$ for some $J \subset \{1, \ldots, l\}$ of cardinality $l - 1$.

Hence, $J(F) = \{1, \ldots, l\}$ implies $\mu(F) \geq e^{-g(1-\varepsilon_r)l}$ and we need only estimate the measure of $F \in \mathfrak{F}(s)$ with $\mu(F) \geq e^{-g(1-\varepsilon_r)l}$. But each function $-\log \mu(f_i)$, $1 \leq i \leq l$, has integral g and, by the weak law of large numbers, the measure of $F \in \mathfrak{F}(s)$ with $\mu(F) \geq e^{-g(1-\varepsilon_r)l}$ tends to zero as $l \to \infty$.

(ii) If $F \in \mathfrak{F}(s)$ satisfies $\dfrac{\text{card}\, J(F)}{l} < 1 - \dfrac{2g}{|\log\theta|}\varepsilon_r - \dfrac{\log\eta}{l\log\theta}$, then

$$\begin{aligned}\mu(F) &\leq \mu(\tilde{F})\,\theta^{l-\text{card}\,J(F)} \\ &\leq \mu(\tilde{F})\,\theta^{\left(\frac{\log\eta}{\log\theta} - \frac{2g}{\log\theta}\varepsilon_r l\right)} \quad (\text{since } 0 < \theta < 1) \\ &\leq \frac{1}{\eta}\, e^{-g(1-\varepsilon_r)l}\, e^{-2g\varepsilon_r l}\,\eta \quad (\text{since } J(F) \neq \{1, \ldots, l\}) \\ &= e^{-g(1+\varepsilon_r)l}\end{aligned}$$

and (ii) follows by an application of the weak law of large numbers as in (i). //

Let $(X_1, \mathcal{B}_1, m_1, T_1)$ and $(X_2, \mathcal{B}_2, m_2, T_2)$ be Bernoulli processes defined by probability vectors $p = (p_0, \ldots, p_{a-1})$ and $q = (q_0, \ldots, q_{b-1})$ with $p_0 = q_0$ and $h(p) = h(q)$. Observe that the filler entropies of the two processes coincide

$$\text{i.e.}\quad g = g(p) = -\sum_{i=1}^{a-1} \frac{p_i}{1-p_0}\log\left(\frac{p_i}{1-p_0}\right) = g(q) = -\sum_{i=1}^{b-1} \frac{q_i}{1-q_0}\log\left(\frac{q_i}{1-q_0}\right).$$

Now that we have fixed our two Bernoulli processes, we may specify the sequence $0 < N_1 < N_2 < \ldots$ used in defining skeletons. Note that in 12 $l_0(\delta, r)$ and $l_1(\delta, r)$ do not depend on this sequence. Choose $1 > \delta_r > 0$ with $\delta_r \to 0$. For $i = 0, 1$ let $l'_i(\delta_r, r)$ be the maximum of the $l_i(\delta_r, r)$ for the two processes $(X_1, \mathcal{B}_1, m_1, T_1)$ and $(X_2, \mathcal{B}_2, m_2, T_2)$. Choose a sequence $\{L_r\}$ such that $L_r \geq \max\{l'_0(\delta_r, r), l'_1(\delta_r, r)\}$ and $\lim_{r\to\infty} L_r(\varepsilon_{r-1} - \varepsilon_r) = \infty$ (e.g. also require $L_r \geq (\varepsilon_{r-1} - \varepsilon_r)^{-2}$). Then $\lim_{r\to\infty} L_r = \infty$ also. Now use 9(ii) to choose $0 < N_1 < N_2 < \ldots$ corresponding to $L_1, L_2, \ldots$. Observe (from the proof of 9(ii)) that, as $p_0 = q_0$, the sequence $\{N_r\}$ will then satisfy 9(ii) for both $(X_1, \mathcal{B}_1, m_1, T_1)$ and $(X_2, \mathcal{B}_2, m_2, T_2)$.

We shall put

$$\eta_1 = \min_{1\leq i\leq a-1}\left\{\frac{p_i}{1-p_0}\right\},\quad \theta_1 = \max_{1\leq i\leq a-1}\left\{\frac{p_i}{1-p_0}\right\},\quad \eta_2 = \min_{1\leq i\leq b-1}\left\{\frac{q_i}{1-q_0}\right\},$$

$$\theta_2 = \max_{1\le i\le b-1} \{\frac{q_i}{1-q_0}\}.$$

If s is a skeleton, the filler set of s for $(X_1, \mathcal{B}_1, m_1, T_1)$ will be denoted by $\mathfrak{F}_1(s)$ and the filler set for $(X_2, \mathcal{B}_2, m_2, T_2)$ by $\mathfrak{F}_2(s)$ and so on. The filler measure on $\mathfrak{F}_i(s)$ is denoted by $\mu_i(i = 1, 2)$. For each skeleton s we shall define a society R_s. R_s is from $\tilde{\mathfrak{F}}_1(s)$ to $\tilde{\mathfrak{F}}_2(s)$ if $r(s)$ is odd, from $\tilde{\mathfrak{F}}_2(s)$ to $\tilde{\mathfrak{F}}_1(s)$ if $r(s)$ is even. Again, the definition is by induction on $r(s)$:

If s is a skeleton of rank 1, let S_s be the society from $\mathfrak{F}_1(s)$ to $\mathfrak{F}_2(s)$ defined by $S_s(\tilde{F}) = \tilde{\mathfrak{F}}_2(s)$, $\tilde{F} \in \tilde{\mathfrak{F}}_1(s)$. Let $R_s < S_s$ be a society chosen as in 4(iii). Suppose R_s has been defined for all skeletons of rank $r - 1$ and r is even. Let s be a skeleton of rank r and rank decomposition $s = s_1 \times \ldots \times s_j$. For each $i = 1, \ldots, j$ we have a society R_{s_i} from $\tilde{\mathfrak{F}}_1(s_i)$ to $\tilde{\mathfrak{F}}_2(s_i)$. Each $R^{\star}_{s_i}$ is a society from $\tilde{\mathfrak{F}}_2(s_i)$ to $\tilde{\mathfrak{F}}_1(s_i)$ so that $S_s = R^{\star}_{s_1} \times \ldots \times R^{\star}_{s_j} = (R_{s_1} \times \ldots \times R_{s_j})^{\star}$ is a society from $\tilde{\mathfrak{F}}_2(s)$ to $\tilde{\mathfrak{F}}_1(s)$. Since each element of $\overline{\mathfrak{F}}_1(s)$ is a disjoint union of elements of $\tilde{\mathfrak{F}}_1(s)$, we may regard S_s as a society from $\overline{\mathfrak{F}}_2(s)$ to $\tilde{\mathfrak{F}}_1(s)$. Let $R_s < S_s$ be a society from $\overline{\mathfrak{F}}_2(s)$ to $\tilde{\mathfrak{F}}_1(s)$, chosen as in 4(iii). Now regard R_s as a society from $\tilde{\mathfrak{F}}_2(s)$ to $\tilde{\mathfrak{F}}_1(s)$. Since $\overline{\mathfrak{F}}_2(s)$ has fewer elements than $\tilde{\mathfrak{F}}_2(s)$, it is important that the refinement comes before passing to $\tilde{\mathfrak{F}}_2(s)$. This provides the inductive step for proceeding from odd ranks to even. The procedure for passing from even ranks to odd ranks is entirely similar.

Roughly speaking our construction will be as follows:

We use the R_s to define the finitary isomorphism $\phi : X_1 \rightarrow X_2$. For $x = (x_n) \in X_1$, $x_0 \neq 0$, let $F_r(x)$ be the filler of $s_r(x)$ for x. By 9(i), $s_r(x)$ and $F_r(x)$ are defined for almost all x with $x_0 \neq 0$. If $x_0 = 0$, take $(\phi x)_0 = 0$. If $x_0 \neq 0$, we show that (almost surely) there is an even r such that $F_r(x)$ is contained in $R_{s_r(x)}(\overline{G})$ for only one $\overline{G} \in \overline{\mathfrak{F}}_2(s_r(x))$ and $\overline{G}$ has the zero coordinate place of x fixed; we take $(\phi x)_0$ to be this fixed symbol. Shifting, we obtain $\phi : X_1 \rightarrow X_2$. $\psi : X_2 \rightarrow X_1$ is similarly obtained, using odd ranks instead of even. We show $\psi = \phi^{-1}$ and that they satisfy all the requirements.

Put $X_1^{(\infty)} = \{x = (x_n) \in X_1 : x_0 \neq 0\}$. For almost all $x \in X_1^{(\infty)}$, $s_r(x)$ will be as in 9(i), $l_r(x)$ will be its length and $F_r(x)$ will be the filler of $s_r(x)$ obtained from x. We shall distinguish the hole of $s_r(x)$ at which the zero coordinate of x occurs. This hole will be referred to as the <u>zero coordinate place</u> (of x in $s_r(x)$).

Put

$$X_1^{(r)} = \{x \in X_1^{(\infty)} : l_{r'}(x) \geq L_{r'} \text{ for all } r' \geq r\}.$$

The sets $X_1^{(r)}$ are increasing and, by 9(ii), $m_1(X_1^{(r)}) \to m_1(X_1^{(\infty)})$ as $r \to \infty$. For a skeleton s of rank $r > 1$ and length l, a hole $k \in \{1, \ldots, l\}$ and $G = (g_1, \ldots, g_l) \in \mathfrak{F}_2(s)$, we shall say that $\overline{G}$ fixes k (at the symbol g_k) if $g'_k = g_k$ for all $G' = (g'_1, \ldots, g'_l) \in \overline{G}$. It is easy to see from the definition of the equivalence class $\overline{G}$ that $\overline{G}$ fixes k iff $k \in \tilde{J}(G)$.

Let r be even. Let s_{r-1} and s_r be skeletons of ranks $r-1$ and r such that s_{r-1} is a subskeleton of s_r and such that they have lengths $l_{r-1} \geq L_{r-1}$ and $l_r \geq L_r$. By the definition of the society R_{s_r}, fewer than $\text{card}(\overline{\mathfrak{F}}_2(s_r))$ elements of $\tilde{\mathfrak{F}}_1(s_r)$ belong to the image of more than one element of $\overline{\mathfrak{F}}_2(s_r)$. Thus, by 12(i), the measure of the set of $F \in \mathfrak{F}_1(s_r)$ for which there exists no unique $\overline{G} \in \overline{\mathfrak{F}}_2(s_r)$ such that $F \in R_{s_r}(\overline{G})$, is less than

$$\delta_r + \text{card}(\overline{\mathfrak{F}}_2(s_r)) \frac{1}{\eta_1} e^{-g(1-\varepsilon_r) l_r}$$

$$\leq \delta_r + \frac{1}{\eta_1} e^{g(1-\varepsilon_{r-1}) l_r} e^{-g(1-\varepsilon_r) l_r} \qquad \text{(by 11(ii))}$$

$$\leq \delta_r + \frac{1}{\eta_1} e^{-g(\varepsilon_{r-1} - \varepsilon_r) L_r} = \xi_r .$$

Therefore, the conditional measure of those $x \in X_1(s_{r-1}, s_r) = \{x \in X_1^{(\infty)} : s_{r-1}(x) = s_{r-1}, s_r(x) = s_r\}$ such that $F_r(x) \notin R_{s_r}(\overline{G})$ for some unique $\overline{G} = \overline{G}(x) \in \overline{\mathfrak{F}}_2(s_r)$ is bounded by ξ_r. Note that $\xi_r \to 0$ and $r \to \infty$.

Now let $F \in \mathfrak{F}_1(s_r)$, $k \in \{1, \ldots, l_{r-1}\}$ and observe the following consequence of the invariance of the measure m_1 : Given $x \in X_1(s_{r-1}, s_r)$ and $F_r(x) = F$, the conditional measure that the zero coordinate place of x is at k is simply $\frac{1}{l_{r-1}}$. It follows that, for any $\overline{G} \in \overline{\mathfrak{F}}_2(s_r)$, the conditional measure of $x \in X_1(s_{r-1}, s_r)$ such that $F_r(x) \in R_{s_r}(\overline{G})$ and $\overline{G}$ fixes the zero coordinate place of x is

$$\frac{1}{l_{r-1}} \{\text{card } J(G|_{s_{r-1}})\} \mu_1(R_{s_r}(\overline{G})) \geq \frac{1}{l_{r-1}} \{\text{card } J(G|_{s_{r-1}})\} \mu_2(\overline{G}),$$

where $G|_{s_{r-1}}$ denotes the subfiller of G for s_{r-1}. By 12(ii) we deduce that the conditional measure of $x \in X_1(s_{r-1}, s_r)$ such that $F_r(x) \in R_{s_r}(\overline{G})$ for some $\overline{G}$

which fixes the zero coordinate place (of x) is greater than or equal to

$$(1 - \frac{2g}{|\log \theta_2|}\varepsilon_{r-1} - \frac{\log \eta_2}{l_{r-1}\log \theta_2})(1 - \delta_{r-1})$$

$$\geq 1 - \frac{2g}{|\log \theta_2|}\varepsilon_{r-1} - \frac{\log \eta_2}{L_{r-1}\log \theta_2} - \delta_{r-1}.$$

In other words, the conditional measure of $x \in X_1(s_{r-1}, s_r)$ such that $F_r(x) \notin R_{s_r}(\overline{G})$ for all $\overline{G}$ which fix the zero coordinate place of x is less than or equal to

$$\zeta_r = \delta_{r-1} + \frac{2g}{|\log \theta_2|}\varepsilon_{r-1} + \frac{\log \eta_2}{L_{r-1}\log \theta_2} .$$

Note that $\zeta_r \to 0$ as $r \to \infty$.

As s_{r-1} and s_r run through all skeletons of ranks $r - 1$ and r where s_{r-1} is a subskeleton of s_r, $l(s_{r-1}) \geq L_{r-1}$ and $l(s_r) \geq L_r$, the sets $X_1(s_{r-1}, s_r)$ form a countable partition of $\{x \in X_1^{(\infty)} : l_{r-1}(x) \geq L_{r-1}, l_r(x) \geq L_r\} \supset X_1^{(r-1)}$. Consider the m_1-measure of $x \in X_1^{(\infty)}$ such that $F_r(x) \in R_{s_r(x)}(\overline{G})$ for a unique $\overline{G} = \overline{G}(x) \in \mathfrak{F}_2(s_r(x))$ and this unique $\overline{G}(x)$ fixes the zero coordinate place of x. It follows from the estimates of the last two paragraphs that this measure is greater than

$$m_1(X_1^{(r-1)}) - \xi_r - \zeta_r ,$$

which tends to $m_1(X_1^{(\infty)})$. Thus $(\phi x)_0$ can be defined a. e. In fact, given any infinite sequence of even r, we can find for almost all x an infinite subsequence all of which may be used to define $(\phi x)_0$. The next lemma shows that $(\phi x)_0$ does not depend on which r is chosen.

13. Lemma. Let $x \in X_1^{(\infty)}$, let r be even and let $i \in \{1, \ldots, b-1\}$. Suppose $F_r(x) \in R_{s_r(x)}(G)$ implies that $\overline{G} \in \mathfrak{F}_2(s_r(x))$ fixes the zero coordinate of x at the symbol i. Then this statement remains true (for the same symbol i) when r is replaced by r + 2.

13 may be checked by going through the inductive definition of the societies R_s.

We have shown that $(\phi x)_0$ is well-defined for almost all $x \in X_1$. Shifting, we obtain ϕ which will then satisfy $\phi T_1 = T_2 \phi$. For all $x = (x_n) \in X_1$ with $x_0 = 0$ we have $(\phi x)_0 = 0$. If $x_0 \neq 0$ then, almost surely, we can find an even r such that for all $x' \in X_1$ with $s_r(x') = s_r(x)$ and $F_r(x') = F_r(x)$ we have

$(\phi x')_0 = (\phi x)_0$. In other words, ϕ is finitary. It follows that ψ is measurable.

14. Lemma. $m_2 = m_1 \circ \phi^{-1}$.

Proof. It is enough to verify this for cylinders $[i_0, \ldots, i_n]$, $i_0, \ldots, i_n \in \{0, \ldots, b-1\}$. It is not hard to see from the construction of ϕ through societies that $m_1(\phi^{-1}[i_0, \ldots, i_n]) \leq m_2[i_0, \ldots, i_n]$. But $\{\phi^{-1}[i_0, \ldots, i_n] : i_0, \ldots, i_n \in \{0, \ldots, b-1\}\}$ is a partition of X_1 and $\sum_{i_0, \ldots, i_n} m_2[i_0, \ldots, i_n] = 1$, so equality must hold on each cylinder. //

Working with odd ranks instead of even, we obtain in the same way a finitary homomorphism $T_2 \xrightarrow{\psi} T_1$. We complete the proof of 2 by showing $\psi = \phi^{-1}$.

15. Lemma. $\psi = \phi^{-1}$.

Proof. Let $x = (x_n) \in X_1$ and put $\phi x = y = (y_n) \in X_2$. It is sufficient to show $(\psi y)_0 = x_0$. If $x_0 = 0$, this is clear. If $x_0 \neq 0$ then, since $x_n = 0$ iff $y_n = 0$, we have $s_r(x) = s_r(y)$ for all $r \geq 1$. Moreover, if we put $s_r = s_r(x) = s_r(y)$ then in each s_r the zero coordinate places of x and y coincide. Choose even r such that $G_{r-1}(y)$, the filler of s_{r-1} given by y, is contained in $R_{s_{r-1}}(\overline{F})$ for a unique $\overline{F} \in \mathfrak{F}_1(s_{r-1})$ and this $\overline{F}$ fixes the zero coordinate place (at the symbol $(\psi y)_0$). Also insist that $F_r(x)$ is contained in $R_{s_r}(\overline{G})$ for a unique $\overline{G} \in \overline{\mathfrak{F}}_2(s_r)$ and that this $\overline{G}$ fixes the zero coordinate place (at the symbol y_0). Such r may be found since any infinite sequence of even (resp. odd) numbers has an infinite subsequence all of which may be used to define ϕ (resp. ψ). Observe that, by definition, $\overline{F}$ contains $F_{r-1}(\psi y)$ and $\overline{G}$ contains $G_r(y)$. $R^{\star}_{s_{r-1}}(\widetilde{G_{r-1}(y)})$ will consist of the elements of $\tilde{\mathfrak{F}}_1(s_{r-1})$ whose union is $\overline{F}$. Hence S_{s_r}, and R_{s_r}, map $\overline{G_r(y)} = \overline{G}$ to a set whose elements fix the zero coordinate place at $(\psi y)_0$. But $F_r(x) \in R_{s_r}(\overline{G})$. //

According to Keane and Smorodinsky [K. S. 2], entropy is a complete invariant for finitary isomorphisms of all finite state Bernoulli processes (i. e. the restriction to processes with three or more states may be removed from 2) :

16. Lemma [K. S. 2]. If $p = (p_0, p_1)$ is a probability vector, we can find a probability vector $q = (q_0, \ldots, q_{b-1})$ with $b \geq 3$, $h(q) = h(p)$ and $q_0^{k-1} q_1 = p_0^{k-1} p_1$

for some $k \geq 2$.

Proof. Choose q_0 with $\max\{p_0, p_1\} < q_0 < 1$. Since $(\frac{p_0}{q_0})^{k-1} p_1 \to 0$ as $k \to \infty$, we may find $k \geq 2$ such that if $q_1 = (\frac{p_0}{q_0})^{k-1} p_1$ then $q_0 + q_1 < 1$ and $h(q_0, q_1, 1-q_0-q_1) < h(p)$. Now use 5(iii). //

Thus, it suffices to construct a finitary isomorphism between the Bernoulli processes defined by p and q with $h(p) = h(q)$ and $p_0^{k-1} p_1 = q_0^{k-1} q_1$. In [K. S. 2] Keane and Smorodinsky remark that this can be done by adapting the marker method to use the word $0\ldots01$ of length k as a 'marker'.

2. FINITE EXPECTED CODE-LENGTHS [P. 7]

17. Definition. Let ϕ be a finitary homomorphism from a countable state process $(X_1, \mathcal{B}_1, m_1, T_1)$ to another, $(X_2, \mathcal{B}_2, m_2, T_2)$. Let α and β denote the state partitions of the processes. By definition, each $\phi^{-1}B$, $B \in \beta$, may be written as a countable union of α-cylinders. By first expressing $\psi^{-1}B$ as a countable union of disjoint cylinders, then combining these cylinders into possible shorter ones, we see that $\phi^{-1}B$ can be written in a disjoint way as $\phi^{-1}B = \bigcup_n C_n^B$ where C_n^B are α-cylinders. Hence we obtain a partition of X_1 into α-cylinders,

$$(\star) \quad \{C_n^B : B \in \beta, \ \phi^{-1}B = \bigcup_n C_n^B\}.$$

Recall that an α-cylinder has the form $T_1^{l} A_{i_{-l}} \cap \ldots \cap A_{i_0} \cap \ldots \cap T_1^{-k} A_{i_k}$, where $A_{i_{-l}}, \ldots, A_{i_k} \in \alpha$, $l \geq 0$, $k \geq 0$. The length of this cylinder is $l + k + 1$, and its future length is l. At $x \in X_1$, define the code-length $l(x)$ of ϕ (w.r.t. $(\star)$) and the future code-length $f(x)$ of ϕ as follows: Select the cylinder C_n^B of the partition $(\star)$ to which x belongs and let $l(x)$ and $f(x)$ equal the length and future length, respectively, of this α-cylinder. Thus l and f are functions of X_1 into $\{0, 1, 2, \ldots\}$. The expected (future) code-length of ψ is $\int l \, dm_1$ ($\int f \, dm_1$). If ϕ is a finitary isomorphism, we may interchange the roles of the processes and replace ϕ by ϕ^{-1} to obtain the inverse code-length $l'(x)$ of ϕ at $x \in X_2$ and the inverse future code-length $f'(x)$ of ϕ at $x \in X_2$. Then the expected inverse (future) code-length of ϕ is $\int l' \, dm_2$ ($\int f' \, dm_2$).

In [K. S. 1] Keane and Smorodinsky show that for any two Bernoulli processes, one having strictly greater entropy, there is a finitary homomorphism from the one with the greater en' opy to the other. It is remarked at the end of the paper that the homomorphism constructed has finite expected code-length. Thus, it is natural to wonder if theorem 2, obtained by methods similar to those in [K. S. 1], can be strengthened to establish finitary isomorphisms with finite expected code and inverse code-lengths between Bernoulli processes of the same entropy. The aim of this section is to show that this is not possible in general.

Let $(X_i, \mathcal{B}_i, m_i, T_i)$, $i = 1, 2$, be ergodic countable state processes with finite entropies. We shall prove that any finitary isomorphism $T_1 \xrightarrow{\phi} T_2$ which has finite expected code and inverse code-lengths must be quasi-regular. Then, by 2.51, the information cocycles I_{T_1} and $I_{T_2} \circ \phi$ are cohomologous, and the invariants of Chapter II may be used. For instance, the group invariant Λ shows that between the Bernoulli processes defined by $(\frac{1}{4}, \frac{1}{4}, \frac{1}{4}, \frac{1}{4})$ and $(\frac{1}{2}, \frac{1}{8}, \frac{1}{8}, \frac{1}{8}, \frac{1}{8})$ (both have entropy $\log 4$) any finitary isomorphism must have infinite expected code-length, or inverse code-length. Similarly, no finitary isomorphism between any two of the Markov chains defined by the matrices $\begin{pmatrix} p & q \\ p & q \end{pmatrix}$, $\begin{pmatrix} p & q \\ q & p \end{pmatrix}$, $\begin{pmatrix} q & p \\ p & q \end{pmatrix}$ $(0 < p < 1,$ $p \neq \frac{1}{2}$, $q = 1 - p)$ can have finite expected code-lengths.

Let $(X_1, \mathcal{B}_1, m_1, T_1)$ and $(X_2, \mathcal{B}_2, m_2, T_2)$ be countable state processes with state partitions α and β. If $T_1 \xrightarrow{\psi} T_2$ is a finitary code and f is its future code-length, we put $A_n = \{x \in X_1 : f(x) \geq n\}$ and $a_n = m_1(A_n)$ for $n \geq 1$. Then $\int f \, dm_1 = \sum_{n=1}^{\infty} a_n$. Moreover:

18. <u>Lemma [P. 7].</u> <u>For</u> $N \geq 0$, $d(\phi^{-1} \bigvee_{i=0}^{\infty} T_2^{-i}\beta, \bigvee_{j=-N}^{\infty} T_1^{-j}\alpha) \leq 2 \sum_{k=N+1}^{\infty} a_k$.

<u>Proof.</u> Fix $N \geq 0$. Write $\beta = \{B_1, B_2, \ldots\}$ and write each $\phi^{-1}(B_l) = \bigcup_n C_n^l$, the unique minimal disjoint union of α-cylinders determining f. For each $l \geq 1$ let B_l' be the union of those C_n^l whose future length is less than or equal to N. Then $B_l' \subset \phi^{-1}(B_l)$ and $B_l' \in \bigvee_{-N}^{\infty} T_1^{-j}\alpha$. We have $f \leq N$ on $\bigcup_{l=1}^{\infty} B_l'$ and $f > N$ on its complement, which must therefore be A_{N+1}. Hence $\beta' = \{B_1' \cup A_{N+1}, B_2', B_3', \ldots\}$ is a partition in $\bigvee_{-N}^{\infty} T_1^{-j}\alpha$ and

$$d(\phi^{-1}\beta, \bigvee_{-N}^{\infty} T_1^{-j}\alpha) \leq d(\phi^{-1}\beta, \beta') \leq 2a_{N+1}.$$

It follows that for each $i \geq 0$,

$$d(T_1^{-i}\phi^{-1}\beta, \bigvee_{-N}^{\infty} T_1^{-j}\alpha) = d(\phi^{-1}\beta, \bigvee_{-(N+i)}^{\infty} T_1^{-j}\alpha) \leq 2a_{N+i+1}$$

and, by 2.46(iv),

$$d(\bigvee_{i=0}^{n} T_1^{-i}\phi^{-1}\beta, \bigvee_{j=-N}^{\infty} T_1^{-j}\alpha) \leq 2\sum_{i=0}^{n} a_{N+i+1} \leq 2\sum_{k=N+1}^{\infty} a_k .$$

The result follows by letting $n \to \infty$ and using 2.46(vi). //

19. Theorem [P.7]. Let ϕ be a finitary isomorphism between two ergodic countable state processes $(X_i, \mathcal{B}_i, m_i, T_i)$ $(i = 1, 2)$ whose state partitions have finite entropies. If ϕ has finite expected future and inverse future code-lengths then the processes are quasi-regularly isomorphic through ϕ.

Proof. Let α and β denote the state partitions of the two processes, and put $\alpha^- = \bigvee_0^{\infty} T_1^{-i}\alpha$, $\beta^- = \bigvee_0^{\infty} T_2^{-i}\beta$. Since $\int f\, dm_1 = \sum_{n=1}^{\infty} a_n < \infty$, 18 shows that for N large enough, $d(\phi^{-1}\beta^-, T^N\alpha^-) < 2$. Since

$$H(T_1^N\alpha^- | \alpha^-) = H(\bigvee_{i=1}^{N} T_1^i\alpha | \alpha^-) \leq N\, H(\alpha) < \infty ,$$

2.47 shows that $d(T^N\alpha^-, \alpha^-) < 2$ also. It follows from (the proof of) 2.50 that $d(\phi^{-1}\beta^-, \alpha^-) < 2$. Similarly, considering ϕ^{-1} instead of ϕ and using $\int f'dm_2 < \infty$, we obtain $d(\alpha^-, \phi^{-1}\beta^-) = d(\phi\alpha^-, \beta^-) < 2$. //

Since $f \leq l$ and $f' \leq l'$, the word 'future' may be deleted from the statement of 19.

If $(X, \mathcal{B}, m, T)$ is the Markov chain defined by the matrix P with left invariant probability vector p, $pP = p$, its inverse (shift) is the Markov chain $(X^\star, \mathcal{B}^\star, m^\star, T^\star)$ defined by the matrix $P^\star$ with $P^\star(i, j) = p(j)\,P(j, i)/p(i)$. Note that $pP^\star = p$. $(X, \mathcal{B}, m, T)$ is finitarily isomorphic to $(X^\star, \mathcal{B}^\star, m^\star, T^\star)$ by an isomorphism which has finite expected code-lengths:

20. Exercise (R. Butler and W. Parry). Suppose $(X, \mathcal{B}, m, T)$ has state space $\{0, 1, \ldots, k-1\}$. Define $\phi : X \to X^\star$ by requiring

$$\phi([0, i_1, i_2, \ldots, i_l, 0]^m) = [0, i_l, i_{l-1}, \ldots, i_1, 0]^m$$

for all cylinders $[0, i_1, \ldots, i_l, 0]^m$ with $i_1, \ldots, i_l \in \{1, \ldots, k-1\}$. In this construction, we are using 0 as a 'marker' (i.e. ϕ maps zeros to zeros), and we are reversing the words between zeros. Show that ϕ is a finitary isomorphism and that it has finite expected code-lengths.

CHAPTER IV: BLOCK-CODES

1. CONTINUITY AND BLOCK-CODES

1. Definition. Let $(X_i, \mathcal{B}_i, m_i, T_i)$ be finite state processes with state partitions α_i $(i = 1, 2)$. A homomorphism $T_1 \xrightarrow{\phi} T_2$ is called a block-homomorphism or a block-code if there exists $p \in \mathbb{N}$ such that $\bigvee_{i=-p}^{p} T_1^i \alpha_1 \geq \phi^{-1} \alpha_2$ i.e. if each $\phi^{-1} A$ $(A \in \alpha_2)$ can be expressed as a union of sets in $\bigvee_{i=-p}^{p} T_1^i \alpha_1$. ϕ is called a block-isomorphism or a faithful block-code if it is an isomorphism and both ϕ, ϕ^{-1} are block codes.

We recall our assumption that all finite state processes are reduced; the truth of the following result depends on this.

2. Proposition. If $\phi : X_1 \to X_2$ is a block-homomorphism between two finite state processes $(X_1, \mathcal{B}_1, m_1, T_1)$ and $(X_2, \mathcal{B}_2, m_2, T_2)$ then there exists a continuous measure-preserving surjection $\phi' : X_1 \to X_2$ such that $\phi' T_1 = T_2 \phi'$ and $\phi' = \phi$ a.e.

Proof. It is easy to see that $\phi^{-1} C$ is a finite union (a.e.) of cylinders when $C \subset X_2$ is a cylinder. Hence for each closed-open set $U_2 \subset X_2$, $\phi^{-1} U_2 = U_1$ a.e. for some closed-open $U_1 \subset X_1$. U_1 is uniquely determined as m_1 is positive on non-empty open sets; we put $\psi U_2 = U_1$. ψ is thus a map from the closed-open subsets of X_2 into the closed-open subsets of X_1 which preserves finite unions and intersections Let $x \in X_1$ and let $C_n \in \bigvee_{i=-n}^{n} T_2^i \alpha_2$ be such that $x \in \psi(C_n)$. As $x \in \bigcap_{n \in \mathbb{N}} \psi(C_n)$, $\{\psi(C_n)\}$ and, hence, $\{C_n\}$ have the finite intersection property. Therefore $\bigcap_{n \in \mathbb{N}} C_n$ is not empty by compactness and, since α_2 is a topological generator, $\bigcap_{n \in \mathbb{N}} C_n$ is a point; we take $\phi' x$ to be this point. It is easy to see that ϕ' is continuous and $\phi' T_1 = T_2 \phi'$. Moreover $\phi'^{-1} B = \phi^{-1} B$ a.e. for each closed-open $B \subset X_2$ and, hence, for each $B \in \mathcal{B}_2$ so that $\phi' = \phi$ a.e. and ϕ' is measure-preserving.

Finally, ϕ' is surjective as $\phi'(X_1)$ is a compact (closed) set of measure 1. //

Using the fact that the closed-open sets of a finite state process are finite unions of cylinders it is easily seen that a continuous homomorphism is a block-code. Thus, according to 2 the terms block-code and continuous homomorphism are (essentially) synonymous. It should also be clear from 2 that block-isomorphism and measure-preserving homeomorphism are synonymous terms:

3. Corollary. If $\phi : X_1 \to X_2$ is a block-isomorphism between finite state processes $(X_1, \mathcal{B}_1, m_1, T_1)$ and $(X_2, \mathcal{B}_2, m_2, T_2)$ then there exists a measure-preserving homeomorphism $\phi' : X_1 \to X_2$ such that $\phi' T_1 = T_2 \phi'$ and $\phi' = \phi$ a. e.

4. Definition. If $(X, \mathcal{B}, m, T)$ is a finite state process, we define the winding numbers groups of T, $W(T)$, to be the group generated by the set $\{m(C) : C \text{ is a cylinder}\}$. Evidently,

$$W(T) = \{\int f\, dm : f : X \to \mathbb{Z} \text{ and } f \text{ is continuous}\}.$$

It is easy to see that if $T_1 \overset{\phi}{\to} T_2$ is a block-code then $W(T_2) \subset W(T_1)$. Therefore the winding numbers group is an invariant of block-isomorphism.

5. Exercise. If T_1, T_2 are the Markov automorphisms defined by $\begin{pmatrix} p & q \\ q & p \end{pmatrix}$, $\begin{pmatrix} p & q \\ p & q \end{pmatrix}$ respectively, check that

$$W(T_1) = \text{Group}\{\tfrac{1}{2} p^n q^m : n, m \in \mathbb{N}\}$$

$$W(T_2) = \text{Group}\{p^n q^m : n, m \in \mathbb{N}\}.$$

This indicates that T_2 could be a factor of T_1 by a map which is at most 2-to-1. Show that this is the case.

Evidently block-isomorphisms are regular and the invariants of Chapter II are valid for block-isomorphisms. For instance the Markov chains defined by the matrices $\begin{pmatrix} p & q \\ p & q \end{pmatrix}$, $\begin{pmatrix} p & q \\ q & p \end{pmatrix}$, $\begin{pmatrix} q & p \\ p & q \end{pmatrix}$ are not block-isomorphic. Moreover, the analogue of 2.31 is valid for block-isomorphism.

In the next section we shall consider bounded-to-one block-codes (i.e. block-codes ϕ for which there is a constant K such that $|\phi^{-1}(x)| \leq K$ for all x) and show that, with the exception of the Λ-invariant, all invariants of Chapter II can be extended to bounded-to-one codes. That the group Λ is not in general invariant under bounded-to-one codes may be seen from 5.

2. BOUNDED-TO-ONE CODES

We first prove a well-known lemma.

6. Lemma. If $T_1 \xrightarrow{\phi} T_2$ is a bounded-to-one homomorphism between finite state processes $(X_i, \mathcal{B}_i, m_i, T_i)$ $(i = 1, 2)$ then $h(T_1) = h(T_2)$.

Proof. It is clear that $h(T_2) \leq h(T_1)$, we show $h(T_2) \geq h(T_1)$. Let α, α' be the state partitions of X_1, X_2 respectively, and put $\beta = \phi^{-1}\alpha'$. By a standard result in the theory of Lebesgue spaces (see [R. 1]), there is a finite partition γ such that $\mathcal{B}_1 = \gamma \vee \phi^{-1}\mathcal{B}_2$ i.e. $\bigvee_{-\infty}^{\infty} T_1^i\alpha = \gamma \vee \bigvee_{-\infty}^{\infty} T_1^i\beta$. Putting $\beta^- = \bigvee_{i=1}^{\infty} T_1^{-i}\beta$, $\gamma^- = \bigvee_{i=1}^{\infty} T_1^{-i}\gamma$ we have $\gamma^- \vee T_1^n\beta^- \uparrow \mathcal{B}_1$. Hence, given $\varepsilon > 0$, $H(\gamma|\gamma^- \vee T_1^n\beta^-) < \varepsilon$ for large enough n and

$$h(T_1) = H(\gamma \vee T_1^n\beta|\gamma^- \vee T_1^n\beta^-)$$

$$\leq \varepsilon + H(T_1^n\beta|T_1^n\beta^-) = \varepsilon + h(T_2).$$

Since $\varepsilon > 0$ may be arbitrarily small, $h(T_1) \leq h(T_2)$. //

7. Proposition [P. 5]. If $T_1 \xrightarrow{\phi} T_2$ is a continuous homomorphism (block-code) between Markov chains $(X_i, \mathcal{B}_i, m_i, T_i)$ $(i = 1, 2)$ then ϕ is bounded-to-one iff $h(T_1) = h(T_2)$.

Proof. If ϕ is bounded-to-one then $h(T_1) = h(T_2)$ by 6. Conversely, suppose $h(T_1) = h(T_2)$. Let α, α' be the state partitions of X_1, X_2 respectively, and put $\beta = \phi^{-1}\alpha'$. We may assume without loss of generality that $\alpha \geq \beta$. (Replace α by $\bigvee_{i=-n}^{n} T_1^i\alpha$ if necessary.) Now, writing $\alpha^n = \bigvee_{i=0}^{n} T_1^{-i}\alpha$, $\beta^n = \bigvee_{i=0}^{n} T_1^{-i}\beta$ and using the fact that α, β are Markov,

$$I(\alpha^n|\beta^n) = I(\alpha^n) - I(\beta^n)$$

$$= f + f \circ T_1 + \ldots + f \circ T_1^{n-1} + (I(\alpha) \circ T_1^n - I(\beta) \circ T_1^n)$$

for $f = I(\alpha|T_1^{-1}\alpha) - I(\beta|T_1^{-1}\beta) = I_{T_1} - I_{T_2} \circ \phi$. Therefore

$$0 \leq I(\alpha^n|\beta^n) \leq f + \ldots + f \circ T_1^{n-1} + K$$

for some constant K. Put $F = \liminf_{n\to\infty} (f + f \circ T_1 + \ldots + f \circ T_1^{n-1})$ so that $F \geq -K$. By Fatou's lemma $\int F \, dm_1 \leq \liminf(n \int f \, dm_1) = 0$ since $\int f \, dm_1 = 0$, so F is integrable. Now, given $\varepsilon > 0$, we see by recurrence that for almost all x $F(x) - F(T_1^n x) = f(x) + \ldots + f(T_1^{n-1} x) < \varepsilon$ for infinitely many n. Hence $F = \liminf(f + \ldots + f \circ T^{n-1}) \leq 0$ also holds and we conclude that $I(\alpha^n|\beta^n) \leq f + \ldots + f \circ T^{n-1} + K = F - F \circ T_1^n + K \leq 2K$ for all $n \in \mathbb{N}$. If $A \in \alpha^n$, $B \in \beta^n$ and $A \subset B$ we have by considering the value of $I(\alpha^n|\beta^n)$ on A, $m(A) \geq e^{-2K} m(B)$ so that the number of $A \subset B$, $A \in \alpha^n$ is bounded above by e^{2K}. Since this upper bound is independent of n and of $B \in \beta^n$, ϕ is at most e^{2K}-to-one. //

8. Corollary [C. P.]. If (X_1, T_1) and (X_2, T_2) are topological Markov chains and if $\phi : X_1 \to X_2$ is a continuous surjection with $\phi T_1 = T_2 \phi$ then ϕ is bounded-to-one iff T_1, T_2 have the same topological entropy.

Proof. For $i = 1, 2$ we denote the topological entropy of T_i by $h(T_i)$ and the entropy of T_i with respect to a T_i-invariant probability m on X_i by $h_m(T_i)$.

Suppose ϕ is bounded-to-one. Clearly $h(T_1) \geq h(T_2)$. Let m_1 be the unique Markov measure on X_1 with $h_{m_1}(T_1) = h(T_1)$ (see 2.26). $m_1 \circ \varphi^{-1}$ is T_2-invariant and, by 6,

$$h(T_1) = h_{m_1}(T_1) = h_{m_1 \circ \phi^{-1}}(T_2) \leq h(T_2).$$

Conversely, if $h(T_1) = h(T_2)$ let m_2 be the unique Markov measure on X_2 with $h_{m_2}(T_2) = h(T_2)$. Using the exercise to follow, find a T_1-invariant probability m_1 such that $m_1 \circ \phi^{-1} = m_2$. Now

$$h_{m_1}(T_1) \geq h_{m_2}(T_2) = h(T_2) = h(T_1)$$

and, by the uniqueness of maximal measures, m_1 is Markov. Considering ϕ as a continuous homomorphism between (X_1, T_1, m_1) and (X_2, T_2, m_2), 7 shows that ϕ must be bounded-to-one. //

9. Exercise. Suppose $T_i : X_i \to X_i$ are homeomorphisms of compact metric spaces $(i = 1, 2)$ and let $\phi : X_1 \to X_2$ be a continuous surjection satisfying

$\phi T_1 = T_2\phi$. Let m_2 be a T_2-invariant probability. Note that the map $f \mapsto f \circ \phi : C(X_2) \to C(X_1)$ is injective and use the Hahn-Banach theorem to show that there is a T_1-invariant probability m_1 with $m_1 \circ \phi^{-1} = m_2$.

10. Proposition. Let $(X_1, \mathcal{B}_1, m_1, T_1)$ and $(X_2, \mathcal{B}_2, m_2, T_2)$ be Markov chains of the same entropy. If $\phi : X_1 \to X_2$ is a continuous surjection with $\phi T_1 = T_2 \phi$ then ϕ is measure-preserving iff

$$I_{T_1} = I_{T_2} \circ \phi + g \circ T_1 - g$$

for some continuous g.

Proof. Suppose ϕ is measure-preserving. Pick p such that $\beta = \phi^{-1}\alpha' \le \bigvee_{-p}^{p} T_1^{i}\alpha$ where α, α' are the state partitions. Put $\alpha^- = \bigvee_0^\infty T_1^{-i}\alpha$, $\beta^- = \bigvee_0^\infty T_1^{-i}\beta$. In order to apply 2.7 we must show that $I(\alpha^- | T_1^{-p}\beta^-)$ is finite. We see from 7 that ϕ is bounded-to-one or, equivalently, that φ^{-1} defines a bounded-to-one map between cylinders. It follows that we have a bounded-to-one map on the one-sided level and that there exists a finite partition γ such that $\gamma \vee \beta^- = \alpha^-$. Now

$$\begin{aligned} &H(\alpha^- | T_1^{-p}\beta^-) \\ &\quad \le H(\alpha^- \vee \gamma \vee \bigvee_0^{p-1} T_1^{-i}\beta | T_1^{-p}\beta^-) \\ &\quad \le H(\gamma) + pH(\beta) + H(\alpha^- | \gamma \vee \beta^-) = H(\gamma) + pH(\beta). \end{aligned}$$

Hence $I(\alpha^- | T_1^{-p}\beta^-)$ is finite and we may apply 2.7 to obtain

$$I_{T_1} = I_{T_2} \circ \phi + g \circ T_1 - g$$

for some finite g. However, as ϕ is a block-code, $I_{T_1} - I_{T_2} \circ \phi$ is a function of finitely many coordinates and, by 2.42, g is a function of finitely many coordinates and therefore continuous.

Conversely if $I_{m_1} = I_{m_2} \circ \phi + g \circ T_1 - g$ for some continuous g, 2.24 implies that

$$\int I_{m_1}\, dm_1 = \int I_{m_2} \circ \phi\, dm_1$$

$$= \int I_{m_2} \, d(m_1 \circ \phi^{-1}) \geq \int I_{m_1 \circ \phi^{-1}} \, d(m_1 \circ \phi^{-1}) = \int I_{m_1} \, dm_1 .$$

Therefore $\int I_{m_2} \, d(m_1 \circ \phi^{-1}) = \int I_{m_1 \circ \phi^{-1}} \, d(m_1 \circ \phi^{-1})$ and the uniqueness in 2.24 shows that $m_1 \circ \phi^{-1} = m_2$. //

With the exception of the group Λ, all invariants of Chapter II depend only on the cocycle-coboundary equation holding, sometimes with restrictions on the coboundary. The coboundary in the cocycle-coboundary equation given by 10 satisfies all these restrictions. Hence all invariants of Chapter II other than Λ are valid for block-codes between Markov chains of the same entropy. In particular, the following holds for Bernoulli processes:

11. Proposition. Between Bernoulli processes given by the probability vectors p and q of the same entropy, there are no block-codes unless q may be obtained from p by a permutation.

11 appears in [T] and was also proved jointly by A. del Junco, M. Keane, B. Kitchens, B. Marcus and L. Swanson.

3. SUSPENSIONS AND WINDING NUMBERS

In this section we justify the term "winding numbers" of definition 4. Fix a finite state process $(X, \mathcal{B}, m, T)$.

12. Definition. Let $\tilde{X}$ be the topological space obtained from $X \times [0, 1]$ by identifying $(x, 1)$ and $(Tx, 0)$ for each $x \in X$. Give $\tilde{X}$ the (quotient) σ-algebra $\tilde{\mathcal{B}}$ and measure $\tilde{m}$ obtained from the σ-algebra $\mathcal{B} \times \mathcal{B}_0$, where $\mathcal{B}_0$ is the Borel σ-algebra of $[0, 1]$, and from the product measure of m with Lebesgue measure. Let $\{T_t : t \in \mathbb{R}\}$ be the flow obtained from: $T_t(x, y) = (x, y+t)$ if $0 \leq y+t < 1$. $(\tilde{X}, \tilde{\mathcal{B}}, \tilde{m}, T_t)$, or simply $\{T_t\}$, is called the suspension of $(X, \mathcal{B}, m, T)$ (or of T).

$\tilde{X}$ is compact and metrizable and $\tilde{\mathcal{B}}$ is the Borel σ-algebra of $\tilde{X}$. $\{T_t\}$ is a continuous one-parameter group of homeomorphisms which may be described more precisely as

$$T_t(x, y) = \begin{cases} (T^n x, \; y+t-n), & 0 \leq y < 1-(t-n) \\ (T^{n+1} x, \; y+t-n-1), & 1-(t-n) \leq y < 1 \end{cases} \quad \text{when } t \in [n, n+1) .$$

It should be clear from this that $\{T_t\}$ preserves $\tilde{m}$. It is not difficult to check that $\{T_t\}$ is ergodic when and only when T is ergodic.

13. Definition. The Bruschlinsky group $Br(Y)$ of a compact metric space Y is the group of all continuous maps to the circle K modulo the subgroup of functions homotopic to a constant map.

It can be shown that $Br(Y)$ is isomorphic to $H^1(Y, \mathbb{Z})$, the first Čech cohomology group of Y; henceforth we shall identify these two groups.

14. Lemma. *Each continuous function* $f : \tilde{X} \to K$ *is homotopic to a map of the form* $g(x, y) = \exp(2\pi i M(x) y)$ *where* $M : X \to \mathbb{Z}$ *is continuous.*

Proof. We can write $f(x, 0) = \exp(2\pi i r(x))$ for some continuous $r : X \to \mathbb{R}$, since X is zero-dimensional. Put

$$h(x, y) = \exp[2\pi i(r(x)(1 - y) + r(Tx) y)].$$

h is homotopic to a constant map and $h(x, 0) = f(x, 0)$. Therefore the map $g_1(x, y) = f(x, y)/h(x, y)$ is homotopic to f and $g_1(x, 0) = g_1(Tx, 1) = 1$ for all $x \in X$. For each $x \in X$ let $M(x)$ be the number of times the loop $g_1(x, y)$ wraps around K as y increases from 0 to 1. Then $M : X \to \mathbb{Z}$ is continuous and, for each $x \in X$, $g(x, y) = \exp(2\pi i M(x) y)$ wraps around K the same number of times as $g_1(x, y)$. Hence g and g_1 are homotopic. //

We have just seen that each continuous $f : \tilde{X} \to K$ has within its homotopy class a function $\exp(2\pi i M(x) y)$, where $M : X \to \mathbb{Z}$ is continuous. Suppose the continuous function $N : X \to \mathbb{Z}$ is such that $\exp(2\pi i N(x) y)$ is in the same homotopy class. We may write $\exp[2\pi i(M(x) - N(x)) y] = \exp[2\pi i r(x, y)]$ for some continuous $r : \tilde{X} \to \mathbb{R}$ with $r(x, 0) \in \mathbb{Z}$ for all $x \in X$. Then $(M(x) - N(x)) y = r(x, y) + P(x, y)$, where $P(x, y) \in \mathbb{Z}$. But, for each $x \in X$, $P(x, y)$ is continuous in the range $0 \le y < 1$ so that $P(x, y) = P(x, 0)$ for all $0 \le y < 1$. Write $P(x) = P(x, 0)$. Now,

$$P(Tx) = -r(Tx, 0) = -r(x, 1) = P(x) - (M(x) - N(x))$$

i.e. $N(x) - M(x) = P(Tx) - P(x)$. We have proved

15. Theorem. $H^1(\tilde{X}, \mathbb{Z})$ *is isomorphic to the group of continuous maps*

$M : X \to \mathbb{Z}$ modulo those of the form $P \circ T - P$, $P : X \to \mathbb{Z}$ continuous.

Suppose $\{S_t\}$ is a one-parameter flow on a compact metric space Y with invariant probability μ. Recall that $f : Y \to K$ is said to be differentiable with respect to the flow at $y \in Y$ if the limit $\lim_{t \to 0} \frac{1}{t}(f(S_t y) - f(y)) = f'(y)$ exists. If $f : Y \to K$ is continuously differentiable put $W_\mu(f) = \frac{1}{2\pi i} \int \frac{f'(y)}{f(y)} d\mu(y)$. We denote the homotopy class of continuous $f : Y \to K$ by $[f]$. Schwartzman [S] showed that for each continuous $f : Y \to K$ there is a continuously differentiable $g \in [f]$ and that $W_\mu(g) = 0$ if the continuously differentiable function g is homotopic to a constant. He used this to define the winding numbers homomorphism:

16. Definition. If $\{S_t\}$ is a one-parameter flow on a compact metric space with invariant probability μ, the winding numbers homomorphism with respect to μ, $W_\mu : H^1(Y, \mathbb{Z}) \to \mathbb{C}$, is defined unambiguously by $W_\mu([f]) = W_\mu(g)$ where $g \in [f]$ is continuously differentiable. The image of this homomorphism is called the winding numbers group of the flow with respect to μ.

Definitions 4 and 16 are related by:

17. Theorem. If $(X, \mathcal{B}, m, T)$ is a finite state process then $W(T)$ is the winding numbers group of its suspension $\{T_t\}$ with respect to $\tilde{m}$.

Proof. Each continuous $f : \tilde{X} \to K$ is, by 14, homotopic to a map of the form $\exp 2\pi i M(x) y$ with $M : X \to \mathbb{Z}$ continuous. Such a map is differentiable with respect to $\{T_t\}$ everywhere except, perhaps, at $(x, 0)$. Nevertheless, it is not difficult to show that

$$W_\mu([f]) = \frac{1}{2\pi i} \int_X \int_0^1 \frac{[\exp 2\pi i M(x) y]'}{\exp 2\pi i M(x) y} \, dy \, dm$$

$$= \int_X \int_0^1 M(x) \, dy \, dm = \int M \, dm.$$

Notice that continuous maps from X to Z are precisely integral linear combinations of characteristic functions of cylinders so that the winding numbers group of $\{T_t\}$ is the group generated by the set $\{m(C) : C \text{ is a cylinder}\}$, which is by definition $W(T)$. //

The spaces X of finite state processes are zero-dimensional compact metric spaces. When the process is ergodic and X (reduced) is infinite this means that X is homeomorphic to a Cantor set. Hence, for the interesting cases there is no direct way of distinguishing between the spaces X. By suspending we obtain a canonical homomorphism from the first cohomology group to the reals which may help in distinguishing between processes from the point of view of block-isomorphism. However:

18. Exercise. If T is a finite state process, show that T and T^{-1} have the same winding numbers homomorphism. In particular, $W(T^{-1}) = W(T)$.

The above exercise implies that the winding numbers group is not a complete invariant of block-isomorphism since, as we shall see in Chapter V, there are topological Markov chains which are not "reversible" i.e. which are not topologically conjugate to their inverses.

4. COMPUTATION OF THE FIRST COHOMOLOGY GROUP

Let $(X, \mathcal{B}, m, T)$ be a finite state process. Denote by $C(X, \mathbb{Z})$ the group of continuous maps from X to $\mathbb{Z}$ and by $B(X, \mathbb{Z})$ the subgroup of maps $f \circ T - f$, $f \in C(X, \mathbb{Z})$. We have seen in the last section that the homomorphism

$$C(X, \mathbb{Z})/B(X, \mathbb{Z}) \xrightarrow{\int(\ )dm} \mathbb{R}$$

is an invariant of block-isomorphism. We now show that $H^1(\tilde{X}, \mathbb{Z})$, by itself, cannot be used for distinguishing between non-trivial Markov chains.

19. Theorem. *If $(X, \mathcal{B}, m, T)$ is a Markov chain such that X is infinite then $H^1(\tilde{X}, \mathbb{Z}) \simeq C(X, \mathbb{Z})/B(X, \mathbb{Z})$ is a free Abelian group with a countable infinity of generators.*

Proof. Let F denote the set of functions in $C(X, \mathbb{Z})$ which are functions of the "past" i.e. measurable with respect to $\bigvee_{i=0}^{\infty} T^{-i}\alpha$ where α is the state partition. A function $f \in C(X, \mathbb{Z})$ depends on finitely many coordinates and since $f, f \circ T, f \circ T^2, \ldots$ are all cohomologous we see that $C(X, \mathbb{Z})/B(X, \mathbb{Z}) \simeq F/\Delta F$ where $\Delta f = f \circ T - f$ and $\Delta F = \{\Delta f : f \in F\}$. Write $F = \bigcup_{n=0}^{\infty} F_n$ where $F_n \subset C(X, \mathbb{Z})$ is the set of functions measurable with respect to $\bigvee_{i=0}^{n} T^{-i}\alpha$. As $f \in F_n$ implies $\Delta f \in F_{n+1}$, we have

$$F_0 \underset{\Delta}{\overset{i}{\rightleftarrows}} F_1 \underset{\Delta}{\overset{i}{\rightleftarrows}} F_2 \underset{\Delta}{\overset{i}{\rightleftarrows}} :::$$

where i is inclusion. Thus we have a directed sequence

$$F_1/\Delta F_0 \xrightarrow{j} F_2/\Delta F_1 \xrightarrow{j} F_3/\Delta F_2 \xrightarrow{j} \cdots$$

where the homomorphisms j are given by $j(f + \Delta F_{n-1}) = f + \Delta F_n$, $f \in F_n$. Each j is injective since, if $f + \Delta F_{n-1} = \Delta F_n$, $f \in F_n$ then $f = g \circ T - g$ with $g \in F_{n-1}$ by 2.42 i.e. $f \in \Delta F_{n-1}$. It is easy to see that $F/\Delta F$ is the direct limit of the above sequence, we now show that $(F_{n+1}/\Delta F_n)/j(F_n/\Delta F_{n-1}) = F_{n+1}/F_n + \Delta F_n$ is torsion free.

Suppose $f \in F_{n+1}$ is such that there is $k \in \mathbb{Z}$ with $kf \in F_n + \Delta F_n$ i.e. $kf = g + h \circ T - h$ for some $g, h \in F_n$. Exponentiating we have

$$\exp(-2\pi i\, g/k) = \exp 2\pi i \frac{h \circ T}{k} / \exp 2\pi i \frac{h}{k}$$

and, by 2.42, $\exp 2\pi i \frac{h}{k}$ is a function of $x_0, \ldots, x_{n-1}$. Put $\exp 2\pi i \frac{h}{k} = \exp 2\pi i \frac{M(x_0, \ldots, x_{n-1})}{k}$, $M : X \to \mathbb{Z}$ so that $h(x_0, \ldots, x_n) - M(x_0, \ldots, x_{n-1}) = kN(x_0, \ldots, x_n)$ where $N : X \to \mathbb{Z}$. Hence, $kf = g + k(N \circ T - N) + M \circ T - M$ and k divides $g + M \circ T - M \in F_n$. It follows that $f = N \circ T - N + f_n$ with $N, f_n \in F_n$, i.e. $f \in F_n + \Delta F_n$, and that $F_{n+1}/F_n + \Delta F_n$ is torsion free. By a similar argument the groups $F_n/\Delta F_{n-1}$ are also torsion free.

We have seen that $F/\Delta F$ is the direct limit of a sequence of finitely generated free Abelian groups whose successive quotients are torsion free. It follows that $F/\Delta F$ is free Abelian with at most countably many generators. If we let θ_n be the number of allowable words $x_0 \ldots x_n$ of length $n + 1$ then the rank of $F_n/\Delta F_{n-1}$ is $\theta_n - \theta_{n-1} + 1$ since Ker Δ consists of constant functions. Clearly $\theta_n - \theta_{n-1} + 1 \to \infty$, and $F/\Delta F$ is infinitely generated. //

20. <u>Exercise.</u> Let T be an ergodic translation of a compact metric Abelian group X. Compute, in terms of the character group of X, $H^1(\tilde{X}, \mathbb{Z})$ where $\tilde{X}$ is the suspension space of X with respect to T.

CHAPTER V: CLASSIFICATIONS OF TOPOLOGICAL MARKOV CHAINS

This chapter is concerned with various classifications of topological Markov chains. We shall often be dealing simultaneously with several topological Markov chains, and it is convenient to denote a topological Markov chain and its defining matrix by the same symbol. For a topological Markov chain S, h(S) denotes its topological entropy. It should be clear from Section 4 of Chapter II that $h(S) = \log \beta$ where β is the maximum eigenvalue of the defining matrix S.

1. FINITE EQUIVALENCE

1. Definition. Two topological Markov chains (X_1, S_1) and (X_2, S_2) are said to be finitely equivalent if there is a topological Markov chain (Y, T) and bounded-to-one continuous surjections $\phi_1 : Y \to X_1$, $\phi_2 : Y \to X_2$ such that $\phi_1 T = S_1 \phi_1$, $\phi_2 T = S_2 \phi_2$.

Finite equivalence may be pictured as

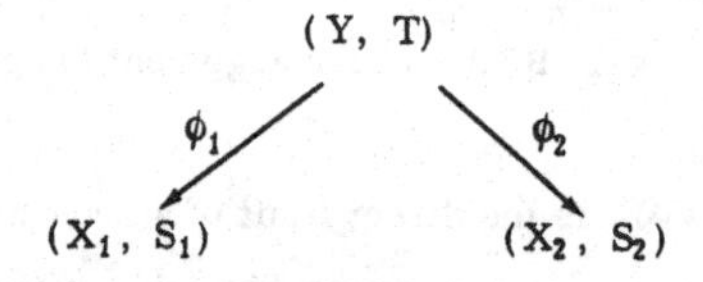

If (X_1, S_1) and (X_2, S_2) are finitely equivalent then, by 4.8, $h(S_1) = h(S_2)$ and the matrices S_1, S_2 have the same maximum eigenvalue. We shall prove that the converse is also true. That finite equivalence is an equivalence relation is an immediate consequence of the converse.

We shall use amalgamation and division matrices to construct continuous semi-conjugacies of topological Markov chains:

2. Definition [W']. A rectangular 0-1 matrix is called a division matrix if its rows are non-trivial and each column contains exactly one non-zero entry. A 0-1 matrix is called an amalgamation matrix if its transpose is a division matrix.

Throughout this section the symbols A, D with or without embellishments will denote amalgamation and division matrices, respectively.

Note that the product of two division (resp. amalgamation) matrices is a division (resp. amalgamation) matrix. A matrix which is both division and amalgamation is a permutation matrix.

3. Proposition. *Let (X, S) and (Z_1, W_1) be topological Markov chains. If their defining matrices satisfy $A_1 S = W_1 A_1$ for some amalgamation matrix A_1, then there exists a continuous surjection $\phi : Z_1 \to X$ such that $\phi W_1 = S\phi$.*

Proof. Suppose S is $k \times k$ and W_1 is $l \times l$. We use the amalgamation A_1 to define $\phi_0 : \{1, \ldots, l\} \to \{1, \ldots, k\}$ and then to extend to $\phi : Z_1 \to X$. A_1 is $l \times k$ and given $x_0 \in \{1, \ldots, l\}$ there exists a unique $y_0 \in \{1, \ldots, k\}$ such that $A_1(x_0, y_0) = 1$, put $\phi_0(x_0) = y_0$. Note that, since S is 0-1 and A_1 an amalgamation, $A_1 S = W_1 A_1$ is 0-1. If $W_1(x_0, x_1) = 1$ then $W_1(x_0, x_1) A_1(x_1, \phi_0 x_1) = 1$ so that

$$1 = (W_1 A_1)(x_0, \phi_0 x_1) = (A_1 S)(x_0, \phi_0 x_1) = A_1(x_0, \phi_0 x_0) S(\phi_0 x_0, \phi_0 x_1).$$

Thus, $W_1(x_0, x_1) = 1$ implies $S(\phi_0 x_0, \phi_0 x_1) = 1$ and the map $\phi(x) = \{\phi_0(x_i)\}$, $x = \{x_i\}$ is well defined. It is clear that ϕ is continuous and $\phi W_1 = S\phi$. We complete the proof by showing that ϕ is surjective. If $\phi_0(x_0) = y_0$ and $S(y_0, y_1) = 1$ then $(W_1 A_1)(x_0, y_1) = A_1(x_0, y_0) S(y_0, y_1) = 1$ and we can find x_1 with $W_1(x_0, x_1) = 1$, $\phi_0(x_1) = y_1$. Thus given $y = (y_i)$ we can, for each $n \geq 0$, find $x^{(n)}$ such that $(\phi x^{(n)})_i = y_i$ for all $i \geq -n$. Now any limit point x of $x^{(n)}$ has $\phi x = y$. //

Interchanging the roles of rows and columns in the above proof we obtain

4. Proposition. *Let (Y, T) and (Z_2, W_2) be topological Markov chains. If their defining matrices satisfy $TD_1 = D_1 W_2$ for some division matrix D_1, then there exists a continuous surjection $\psi : Z_2 \to Y$ such that $\psi W_2 = T\psi$.*

In order to establish topological entropy as a complete invariant of finite equivalence, we shall show that if S, T are topological Markov chains with $h(S) = h(T)$ then the hypotheses of 3 and 4 can be satisfied with $(Z_1, W_1) = (Z_2, W_2)$.

Given a non-negative integral $k \times k$ matrix M, we have a graph with k

nodes and $M(i, j)$ paths from i to j. Let Q be the totality of the (directed) paths in the graph, in some fixed order. We may index with Q the columns of a $\times |Q|$ division matrix D with $D(i, q) = 1$ iff q starts at i. Similarly we have a $|Q| \times k$ amalgamation matrix A with $A(q, j) = 1$ iff q terminates at j. Now $M = DA$ and it is not difficult to prove

5. Proposition [W']. If M is a non-trivial non-negative integral matrix then it can be written as $M = DA$ where D is a division matrix and A is an amalgamation. This decomposition into the product of a division matrix with an amalgamation matrix is essentially unique in the sense that if $M = D'A'$ then $D' = DP$, $A' = P^{-1}A$ for some permutation matrix P.

6. Example. If $M = \begin{pmatrix} 0 & 2 \\ 1 & 1 \end{pmatrix}$ then the graph

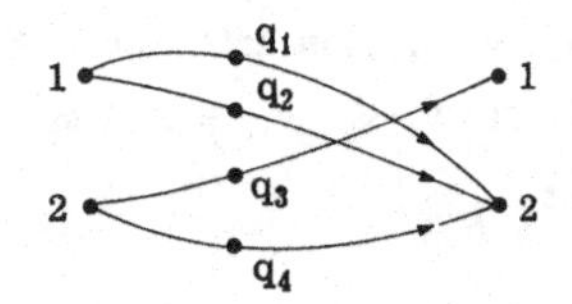

gives the decomposition $M = \begin{pmatrix} 1 & 1 & 0 & 0 \\ 0 & 0 & 1 & 1 \end{pmatrix} \begin{pmatrix} 0 & 1 \\ 0 & 1 \\ 1 & 0 \\ 0 & 1 \end{pmatrix}$.

7. Lemma (Furstenberg). Let S, T be irreducible non-negative integral matrices. S, T have the same maximum eigenvalue iff there exists a strictly positive integral matrix R with $RS = TR$.

Proof. Suppose $RS = TR$ where R is strictly positive. Let β be the maximum eigenvalue of S and let w be a corresponding strictly positive eigenvector, $Sw = \beta w$. Then $T(Rw) = \beta(Rw)$, Rw is a strictly positive vector and the Perron-Frobenius theorem shows that β must be the maximum eigenvalue of T.

Conversely suppose $uS = \beta u$ and $Tv = \beta v$ where u, v are strictly positive row and column vectors respectively. Then

$$T(vu) = \beta(vu) = (vu)S$$

but the strictly positive matrix vu is not in general integral. Nevertheless for every $\varepsilon > 0$ there exists an integer n such that $n(vu) = R + E$ where R is a strictly positive integral matrix and each entry of E has absolute value less than ε. Since $T(nvu) = (nvu)S$ we obtain $TR = RS$ by choosing $\varepsilon > 0$ small enough. //

8. Lemma. *If the non-negative non-trivial integral matrices R, S, T satisfy $RS = TR$ then there exist non-negative non-trivial integral matrices W, D, A such that D is a division matrix, A is an amalgamation and the following diagram commutes:*

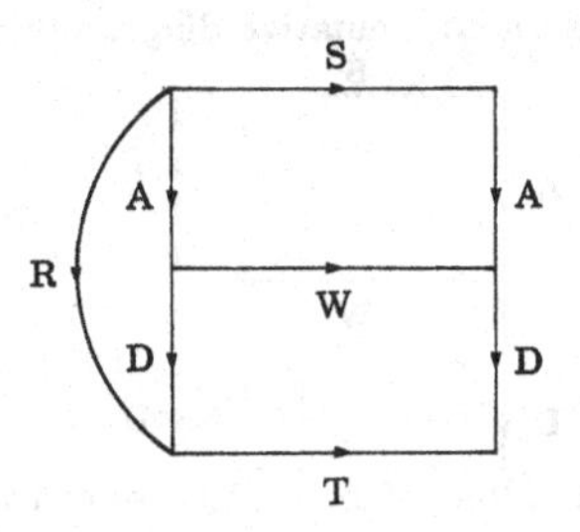

Proof. Use 5 to decompose $R = DA$, $S = D_1A_1$, $T = D_2A_2$ into products of division matrices with amalgamation matrices and rewrite $AD_1 = D'A'$, $A_2D = D''A''$ to obtain the diagram:

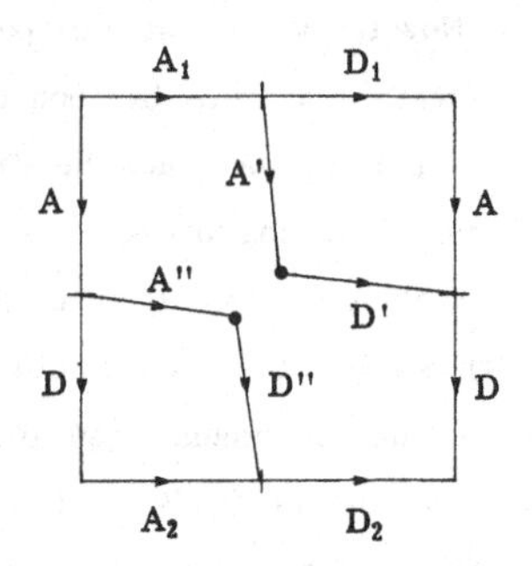

Now $DD'A'A_1$ and $D_2D''A''A$ are two decompositions of $RS = TR$. Replacing D'', A'' by $D''P$, $P^{-1}A''$ where P is a suitable permutation, we may assume $DD' = D_2D''$ and $A'A_1 = A''A$. Defining $W = D'A''$, the lemma is established. //

9. Proposition. *If S, T are irreducible non-negative integral matrices with the same maximum eigenvalue β then there exist an irreducible non-negative integral matrix W_1 with maximum eigenvalue β and division and amalgamation matrices D_1, A_1 such that the following diagram is commutative.*

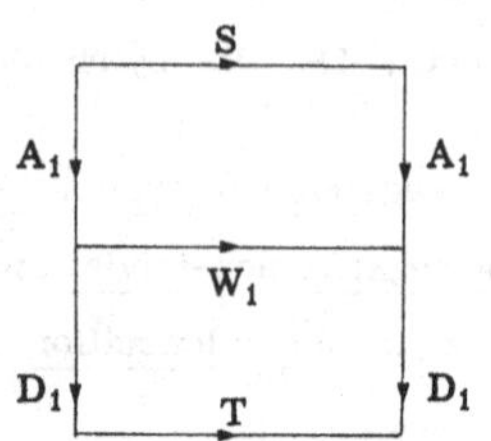

Proof. By 7 and 8 we have a commutative diagram where A is an amalgama-

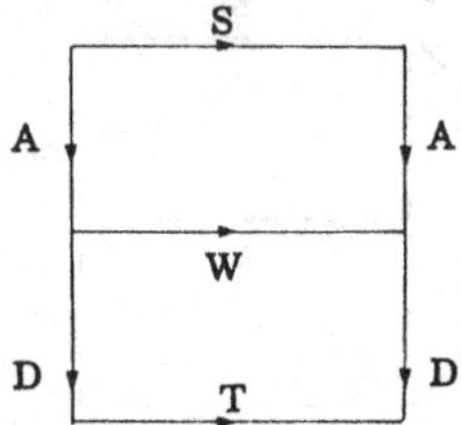

tion and D is a division. Without loss of generality we can write $W = \begin{pmatrix} W_1 & 0 \\ B & W_2 \end{pmatrix}$ with W_1 irreducible. Accordingly, write $A = \begin{pmatrix} A_1 \\ A_2 \end{pmatrix}$. Then from $\begin{pmatrix} A_1 \\ A_2 \end{pmatrix} S = \begin{pmatrix} W_1 & 0 \\ B & W_2 \end{pmatrix} \begin{pmatrix} A_1 \\ A_2 \end{pmatrix}$ we obtain $A_1 S = W_1 A_1$. Choose a strictly positive vector v such that $Sv = \beta v$. Then $W_1(A_1 v) = A_1(Sv) = \beta(A_1 v)$. Since no row of the 0-1 matrix A_1 is trivial, $A_1 v$ is strictly positive and we see that β is also the maximum eigenvalue of W_1. Now let w be a strictly positive row vector with $wW_1 = \beta w$. Then $(wA_1)S = \beta(wA_1)$ and wA_1 is a non-trivial non-negative vector. As β is the maximum eigenvalue of S, wA_1 must be strictly positive and it follows that A_1 has no trivial columns. Recalling that each row of A_1 is also a row of the amalgamation matrix A we conclude that A_1 is an amalgamation matrix.

Now write $D = (D_1, D_2)$ so that $DW = TD$ implies $D_1 W_1 + D_2 B = TD_1$. We know that T and W_1 have the same maximum eigenvalue, β. If u is a strictly positive vector such that $uT = \beta u$, then $(uD_1)W_1 + u(D_2 B) = \beta(uD_1)$. Hence $(uD_1)W_1 \leq \beta(uD_1)$ and, to avoid contradicting the Perron-Frobenius theorem, equality must hold i.e. $u(D_2 B)$ must be zero. It follows that $B = 0$ and $D_1 W_1 = TD_1$. We may now apply our previous reasoning to conclude that D_1 is a division matrix. //

10. Theorem [P5]. Two topological Markov chains are finitely equivalent iff they have the same topological entropy.

Proof. We have already remarked that 4.8 implies that topological entropy is an invariant of finite equivalence of topological Markov chains. For the converse, suppose the topological Markov chains S, T have the same topological entropy. This is equivalent to supposing that the defining 0-1 matrices S, T have the same maximum eigenvalue. From 9 we obtain an irreducible non-negative matrix W_1 (with the same maximum eigenvalue as S, T) and amalgamation and division matrices A_1, D_1 such that $A_1 S = W_1 A_1$, $TD_1 = D_1 W_1$. Since A_1 is an amalgamation and S is 0-1, the matrices $A_1 S = W_1 A_1$ and W_1 must be 0-1. Applying 3 and 4 we find continuous surjections ϕ, ψ such that $\phi W_1 = S\phi$, $\psi W_1 = T\psi$. Since W_1 has the same topological entropy as S, T, 4.8 shows that ϕ, ψ must be bounded-to-one, and the proof is complete. //

2. ALMOST TOPOLOGICAL CONJUGACY AND THE ROAD PROBLEM

11. Definition. Let (X, S) be a topological Markov chain. A Borel set $N \subset X$ is said to be universally null if it is a null set with respect to every ergodic (Borel) measure with support X. Two topological Markov chains (X_1, S_1) and (X_2, S_2) are said to be almost topologically conjugate if there exists a finite equivalence

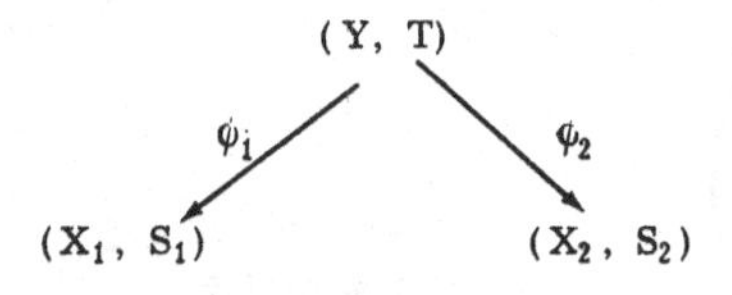

with invariant universally null sets $N_i \subset X_i$ such that the restrictions $\phi_i | Y - \phi_i^{-1}(N_i)$ are injective $(i = 1, 2)$.

The problem of classifying topological Markov chains with respect to almost topological conjugacy arose in the work of Adler and Weiss on automorphisms of the two-dimensional torus [A. W.]. The specific topological Markov chains arising in this connection were classified in [A. W.] but the general problem, of independent interest, remained open. Adler, Goodwyn and Weiss [A. G. W.] later achieved a solution for a subclass. They proved that for aperiodic matrices with integral maximum eigenvalue, their associated topological Markov chains are classified by this eigenvalue. The general problem was solved by Adler and Marcus [A. M.]. We simply state this solution and refer the reader to [A. M.] for the (rather intricate) proof.

12. Theorem [A. M.]. Two topological Markov chains are almost topologically conjugate iff they have the same period and the same topological entropy.

In [A. W.] another problem came up as the result of an attempt to solve the almost topological conjugacy problem for aperiodic 0-1 (defining) matrices with integral maximum eigenvalues (see also [A. G. W.] and [M]). This problem, called the road problem, is still unsolved. We describe the road problem and its relevance to almost topological conjugacy.

Let $n \geq 2$ be an integer and let T be a $k \times k$ aperiodic 0-1 matrix with all its rows sums equal to n. The matrix T gives a graph with k "cities" and n "roads" exiting from each city. Choose a city as the "capital". The road problem for T consists of using $a_1, \ldots, a_n$ to label the roads leaving each city in such a way that there exists a finite word made up of symbols from the set $\{a_1, \ldots, a_n\}$ with the property that, no matter which city you are in, following the entire word will bring you to the capital. Note that, by irreducibility, the solvability of the problem does not depend on the choice of capital.

13. Example. Let

$$T = \begin{pmatrix} 0 & 1 & 1 & 0 & 0 \\ 0 & 0 & 1 & 1 & 0 \\ 0 & 0 & 0 & 1 & 1 \\ 1 & 0 & 0 & 0 & 1 \\ 1 & 1 & 0 & 0 & 0 \end{pmatrix}$$

and let 1 be the capital city.

We label the roads in the graph obtained from T as follows

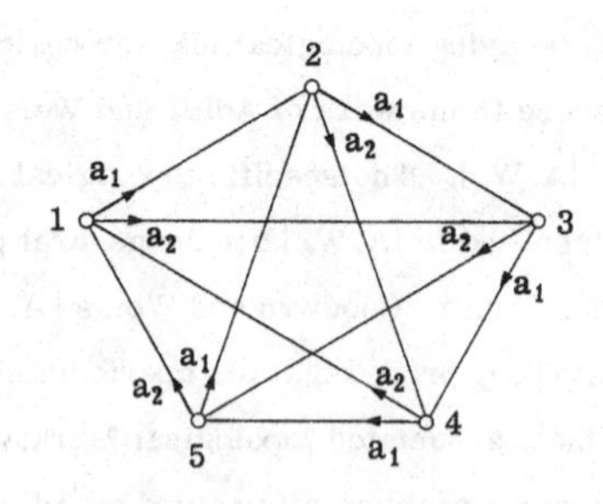

The word $a_1 a_2 a_2$ sends 1, 2, 5 to 1 and 3, 4 to 3 and the word $a_2 a_2 a_1$ sends 1, 3 to 2 so that $a_1 a_2 a_2 a_2 a_2 a_1 a_2 a_2$ sends all cities to 1.

14. Exercise. Show that the road problem is solvable whenever T has non-zero trace.

It is not known if the road problem is solvable for all aperiodic 0-1 matrices with row sums equal to n.

The full n-shift is the topological Markov chain defined by the $n \times n$ matrix of ones. The connection between the road problem and almost topological conjugacy is that if the road problem is solvable for an aperiodic 0-1 matrix T with row sums equal to n, then the solution gives an almost topological conjugacy between the full n-shift and the topological Markov chain T:

Let (Y, T) be the topological Markov chain given by T and let (X, S) be the full n-shift, with $X = \prod_{-\infty}^{\infty} \{a_1, \ldots, a_n\}$. We simply define a map $\phi : Y \to X$ which satisfies all the requirements for an almost topological conjugacy

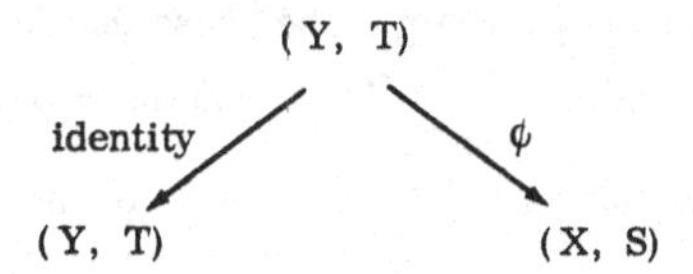

If $T(y_0, y_1) = 1$, define $\phi_0(y_0, y_1) \in \{a_1, \ldots, a_n\}$ according to the label of the road from y_0 to y_1. For $y = (y_i) \in Y$ define $\phi(y)_i = \phi_0(y_i, y_{i+1})$. ϕ amounts to writing a given doubly infinite sequence of transitions between cities as a sequence of road labels. Evidently $\phi T = S\phi$ and ϕ is continuous. Note that for $y = (y_i) \in Y$ and any integer m, y_m and $\phi(y)_i$, $i \geq m$, determine all y_i, $i \geq m$. All the required properties of ϕ follow from this observation, as outlined in the exercise below.

15. Exercise. (i) Use an argument similar to the one in the proof of 3 to show that ϕ is surjective.

(ii) Suppose T is $k \times k$. Show that for any $a = (a_{k_i})_{i \in \mathbb{Z}}$ the inequality $|\phi^{-1}(a)| \leq k$ is valid.

(iii) Let $a = (a_{k_i})_{i \in \mathbb{Z}}$. Show that if $(a_{k_{-i}})_{i \in \mathbb{N}}$ contains the code word solving the road problem infinitely often then $|\phi^{-1}(a)| = 1$.

(iv) Check that the set

$$\{a = (a_{k_i}) \in X : \text{the code word appears only a finite number of times in } (a_{k_{-i}})_{i \in \mathbb{N}}\}$$

is universally null.

3. TOPOLOGICAL CONJUGACY OF TOPOLOGICAL MARKOV CHAINS

In this section we consider topological conjugacy of topological Markov chains and seek necessary and sufficient algebraic conditions on defining matrices.

Recall that two topological Markov chains are topologically conjugate if between them there is a homeomorphism which conjugates the shifts. If we specify the characteristic properties of state partitions which are preserved under topological conjugacy then the problem can be stated as that of determining, for a given topological Markov chain, all partitions with these properties.

16. Definition. Let (X, S) be a topological Markov chain. A partition α of X is called a Markov partition or a Markov generator if it enjoys the following three properties:

(i) α is a finite partition of X into closed-open sets,

(ii) α is a topological generator in the sense that for any sequence of sets $A_{k_i} \in \alpha$, $i \in \mathbb{Z}$, the intersection $\bigcap_{-\infty}^{\infty} S^{-i} A_{k_i}$ contains at most one point,

(iii) if $A_{k_i} \in \alpha$ satisfy $A_{k_i} \cap S^{-1} A_{k_{i+1}} \neq \emptyset$ for all $i \in \mathbb{Z}$ then $\bigcap_{-\infty}^{\infty} S^{-i} A_{k_i} \neq \emptyset$

In the above definition, (ii) and (iii) imply that α determines all points of X. Evidently, the state partition of a topological Markov chain is a Markov partition.

17. Definition. Two non-negative integral matrices S, T are said to be strong shift equivalent (in l steps) if there exist non-negative integral matrices $U_1, \ldots, U_l$ and $V_1, \ldots, V_l$ such that

$$S = U_1 V_1, \quad V_1 U_1 = U_2 V_2, \ldots, V_{l-1} U_{l-1} = U_l V_l, \quad V_l U_l = T.$$

We shall establish Williams's result that topological Markov chains S, T are topologically conjugate iff their defining matrices S, T are strong shift equivalent. For this we need to fix ways of obtaining 0-1 matrices from finite partitions.

Fix a topological Markov chain S. If $\xi = (E_1, E_2, \ldots, E_k)$ and $\eta = (F_1, \ldots, F_l)$ are two (finite) ordered partitions the $k \times l$ zero-one matrix (ξ, η) is defined by

$$(\xi, \eta)(i, j) = \begin{cases} 1 & \text{when } E_i \cap F_j \neq \emptyset \\ 0 & \text{otherwise}. \end{cases}$$

The $k \times l$ zero-one matrix $(\xi, \eta)_S$ is defined by

$$(\xi, \eta)_S(i, j) = \begin{cases} 1 & \text{when } E_i \cap S^{-1}F_j \neq \emptyset \\ 0 & \text{otherwise}. \end{cases}$$

Put $M(\xi) = (\xi, \xi)_S$.

18. Lemma [P. W']. If $\xi \leq \eta \leq \xi \vee S^{-1}\xi$ and if ξ is Markov, then η is Markov and

(i) $M(\xi) = (\xi, \eta)(\eta, \xi)_S$,

(ii) $M(\eta) = (\eta, \xi)_S(\xi, \eta)$.

Proof. Since ξ is Markov, $\xi \vee S^{-1}\xi$ is Markov. It follows easily from $\xi \leq \eta \leq \xi \vee S^{-1}\xi$ that η is also Markov.

Let $E_{i_1}, E_{i_2} \in \xi$ and let $F \in \eta$. Observe that, as $\xi \leq \eta$, $F \cap E_{i_1} \neq \emptyset$ iff $F \subset E_{i_1}$. Similarly, $E_{i_1} \cap S^{-1}E_{i_2} \cap F \neq \emptyset$ iff $E_{i_1} \cap S^{-1}E_{i_2} \subset F$.

For (i), fix $(i_1, i_2) \in \{1, \ldots, k\}^2$. $[(\xi, \eta)(\eta, \xi)_S](i_1, i_2)$ is the number of $F \in \eta$ which intersect both E_{i_1} and $S^{-1}E_{i_2}$. If $E_{i_1} \cap S^{-1}E_{i_2} = \emptyset$, i.e. if $M(\xi)(i_1, i_2) = 0$, then $[(\xi, \eta)(\eta, \xi)_S](i_1, i_2) = 0$ since all $F \in \eta$ with $F \cap E_{i_1} \neq \emptyset$ are contained in E_{i_1}. If $E_{i_1} \cap S^{-1}E_{i_2} \neq \emptyset$, then there exists a unique $F \in \eta$ with $E_{i_1} \cap S^{-1}E_{i_2} \subset F$. Noting that $F' \cap E_{i_1} \neq \emptyset$, $F' \cap S^{-1}E_{i_2} \neq \emptyset$ imply $F' \subset E_{i_1}$, $S^{-1}E_{i_2} \cap E_{i_1} \cap F' = S^{-1}E_{i_2} \cap F' \neq \emptyset$ and $E_{i_1} \cap S^{-1}E_{i_2} \subset F'$, we conclude that $[(\xi, \eta)(\eta, \xi)_S](i_1, i_2) = 1$ whenever $M(\xi)(i_1, i_2) = 1$.

For (ii), fix $(j_1, j_2) \in \{1, \ldots, l\}^2$ and let $(i_1, i_2) \in \{1, \ldots, k\}^2$ be such that $F_{j_1} \subset E_{i_1}$, $F_{j_2} \subset E_{i_2}$. Then $[(\eta, \xi)_S(\xi, \eta)](j_1, j_2)$ is 0 or 1 according as $F_{j_1} \cap S^{-1}E_{i_2}$ is empty or not. Hence, the right hand side is 0 iff $F_{j_1} \cap S^{-1}E_{i_2} = \emptyset$ which implies that $F_{j_1} \cap S^{-1}F_{j_2} = \emptyset$ and this happens iff $M(\eta)(j_1, j_2) = 0$. Suppose $F_{j_1} \cap S^{-1}E_{i_2} \neq \emptyset$. Then $E_{i_1} \cap S^{-1}E_{i_2} \neq \emptyset$. Writing $F_{j_1} = \bigcup_E (E_{i_1} \cap S^{-1}E)$ we see that

$$F_{j_1} \cap S^{-1}F_{j_2} = \bigcup_E (E_{i_1} \cap S^{-1}E \cap S^{-1}F_{j_2}) = E_{i_1} \cap S^{-1}E_{i_2} \cap S^{-1}F_{j_2} = E_{i_1} \cap S^{-1}F_{j_2}.$$

Take any non-empty $E_{i_2} \cap S^{-1}E_{i_3} \subset F_{j_2}$. We have

$$E_{i_1} \cap S^{-1}E_{i_2} \cap S^{-2}E_{i_3} \subset E_{i_1} \cap S^{-1}F_{j_2} = F_{j_1} \cap S^{-1}F_{j_2}$$

where $E_{i_1} \cap S^{-1}E_{i_2} \cap S^{-2}E_{i_3}$ is non-empty by the Markov property. Hence $F_{j_1} \cap S^{-1}F_{j_2} \neq \emptyset$ and $M(\eta)(j_1, j_2) = 1$. //

Let α be a Markov generator for the topological Markov chain (X, S). It is easy to see that the partitions $\alpha^r = \bigvee_{i=0}^{r} S^{-i}\alpha$, $r \geq 0$, are also Markov generators. A topological generator α determines the topology of X, in the following way: the sets of the partitions $\bigvee_{i=k_1}^{k_2} S^{-i}\alpha$, $k_1, k_2 \in \mathbb{Z}$, $k_1 \leq k_2$ (the α-cylinders) form a base for the topology of X. Moreover, the closed-open sets of X are precisely those that can be expressed as finite unions of α-cylinders. It follows that if β is another Markov generator then we can find $p \in \mathbb{N}$ such that $\beta \leq \bigvee_{i=-p}^{p} S^{-i}\alpha$, $\alpha \leq \bigvee_{i=-p}^{p} S^{-i}\beta$. Putting $\gamma = S^{-p}\beta$, $n = 2p$ we have $\gamma \leq \bigvee_{i=0}^{n} S^{-i}\alpha$, $\alpha \leq \bigvee_{i=0}^{n} S^{i}\gamma$. Using $\gamma \leq \bigvee_{i=0}^{n} S^{-i}\alpha$ we obtain

$$(\alpha^{2n})^1 \geq \alpha^{2n} \vee \gamma^{n+1} \geq \alpha^{2n}$$

$$(\alpha^{2n} \vee \gamma^{n+1})^1 \geq \alpha^{2n} \vee \gamma^{n+2} \geq \alpha^{2n} \vee \gamma^{n+1}$$

$$\vdots$$

$$(\alpha^{2n} \vee \gamma^{2n-1})^1 \geq \alpha^{2n} \vee \gamma^{2n} \geq \alpha^{2n} \vee \gamma^{2n-1}$$

Now applying 18 to this sequence of relationships we see that (all partitions involved are Markov and)

$$M(\alpha^{2n}) = U_1 V_1, \ V_1 U_1 = M(\alpha^{2n} \vee \gamma^{n+1}) = U_2 V_2, \ \ldots, \ V_n U_n = M(\alpha^{2n} \vee \gamma^{2n}),$$

for suitable 0-1 matrices $U_1, \ldots, U_n$, $V_1, \ldots, V_n$. In other words, $M(\alpha^{2n})$ and $M(\alpha^{2n} \vee \gamma^{2n})$ are strong shift equivalent in n steps. Interchanging the roles of α and γ, and of the shift S and its inverse, and using $\alpha \leq \bigvee_{i=0}^{n} S^{i}\gamma$, we see that $M(\bigvee_{i=0}^{2n} S^{i}\gamma) = M(\gamma^{-2n}) = M(\gamma^{2n})$ and $M(\bigvee_{i=0}^{2n} S^{i}\gamma \vee \bigvee_{i=0}^{2n} S^{i}\alpha) = M(\gamma^{2n} \vee \alpha^{2n})$ are strong shift equivalent in n steps. Hence $M(\alpha^{2n})$ and $M(\beta^{2n}) = M(\gamma^{2n})$ are strong shift equivalent in $2n$ steps:

19. Proposition. If α, β are two Markov partitions for a topological Markov chain then there exists $n \in \mathbb{N}$ such that the matrices $M(\alpha^{2n})$ and $M(\beta^{2n})$ are strong shift equivalent in $2n$ steps.

20. Theorem ([W'], [P.W']). Two topological Markov chains S, T are topologically conjugate iff their defining matrices S, T are strong shift equivalent.

Proof. Suppose S, T are topologically conjugate by ϕ, $\phi S = T\phi$. Let α, α' be the state partitions of S, T respectively and put $\beta = \phi^{-1}\alpha'$. α, β are Markov partitions for S and we obtain from 19 a positive integer n such that $M(\alpha^{2n})$ and $M(\beta^{2n})$ are strong shift equivalent. It is not hard to see that $M(\alpha^{2n})$ is strong shift equivalent to $M(\alpha)$ and $M(\beta^{2n})$ to $M(\beta) = M(\alpha')$. Therefore $S = M(\alpha)$ and $T = M(\alpha')$ are strong shift equivalent.

For the converse it is sufficient, by the transitivity of topological conjugacy, to prove that if the non-negative matrices S', T', U, V satisfy $S' = UV$, $VU = T'$ and S', T' are irreducible then the topological Markov chains S', T' are topologically conjugate. We do not assume that S', T' are 0-1 since in a strong shift equivalence between 0-1 matrices S, T there is no guarantee that the intermediate matrices are 0-1. Consider the graphs associated with S' and T'. U may be regarded as determining paths from S'-vertices to T'-vertices and V as determining paths from T'-vertices to S'-vertices. Since $S'(i_0, i_1) = \sum_j U(i_0, j)V(j, i_1)$, for each pair of S'-vertices (i_0, i_1) the (U, V)-paths leading from i_0 to i_1 are in bijective correspondence with S'-paths from i_0 to i_1. Fix a bijection for each pair (i_0, i_1). Similarly for each pair of T'-vertices (j_0, j_1) the (V, U)-paths leading from j_0 to j_1 are in bijective correspondence with T'-paths from j_0 to j_1. Fix a bijection for each pair (j_0, j_1) also. Now, given a doubly infinite sequence of S'-paths $(s_n)_{n\in\mathbb{Z}}$ we obtain a sequence of (U, V)-paths

$$\ldots (u_{-1}v_{-1})(u_0v_0)(u_1v_1)(u_2v_2)\ldots$$

where u_nv_n corresponds to s_n. Re-bracketing we get

$$\ldots (v_{-1}u_0)(v_0u_1)(v_1u_2)\ldots$$

which determines a doubly infinite sequence of T'-paths $(t_n)_{n\in\mathbb{Z}}$ in which t_n corresponds to v_nu_{n+1}. It is not hard to check that the map thus obtained is indeed a topological conjugacy between the topological Markov chains S' and T'. //

21. Remark. The relationship between Markov generators and topological conjugacy of topological Markov chains should perhaps be made more explicit. Fix a topological Markov chain S. If the topological Markov chain T is topologically conjugate to S, then the pull back of the state partition of T is a Markov partition of S. Now suppose β is a Markov partition of S. It is well known, and easy to

check, that the topological Markov chain S (defined by an irreducible matrix) is topologically transitive i.e. given open sets $\mathcal{O}$, $\mathcal{O}'$ it is possible to find $n \in \mathbb{N}$ such that $\mathcal{O} \cap S^{-n}\mathcal{O}' \neq \emptyset$. It follows that the matrix $M(\mathcal{P})$ is irreducible. It is easy to produce a topological conjugacy between the topological Markov chains S and $M(\mathcal{P})$ such that the pull back of the state partition of $M(\mathcal{P})$ is $\mathcal{P}$.

Strong shift equivalence has the disadvantage that it is difficult to check. In trying to replace it by a more manageable condition Williams defined shift equivalence.

22. Definition. Two non-negative integral matrices S, T are said to be shift equivalent if there exist non-negative integral matrices U, V and $l \in \mathbb{N}$ such that

$$SU = UT, \ VS = TV, \ UV = S^l \ \text{ and } \ VU = T^l .$$

l is called the lag of the shift equivalence.

23. Proposition [W']. Shift equivalence of non-negative integral matrices is an equivalence relation.

Proof. Reflexivity and symmetry are clear. Suppose non-negative integral matrices R, S, T, U_1, V_1, U_2, V_2 and $m, n \in \mathbb{N}$ satisfy

$$RU_1 = U_1S, \ V_1R = SV_1, \ U_1V_1 = R^m, \ V_1U_1 = S^m ,$$

$$SU_2 = U_2T, \ V_2S = TV_2, \ U_2V_2 = S^n, \ V_2U_2 = T^n .$$

Take $U = U_1U_2$, $V = V_2V_1$. Then

$$RU = UT, \ VR = TV, \ UV = R^{m+n}, \ VU = T^{m+n}$$

and transitivity is verified. //

24. Proposition. If two non-negative integral matrices S, T are strong shift equivalent in l steps then they are shift equivalent with lag l.

Proof. Given a strong shift equivalence

$$S = U_1V_1, \ V_1U_1 = U_2V_2, \ \dots, \ V_{l-1}U_{l-1} = U_lV_l, \ V_lU_l = T$$

we put $U = U_1 U_2 \ldots U_l$, $V = V_l V_{l-1} \ldots V_1$ to obtain a shift equivalence

$$SU = UT, \quad VS = TV, \quad UV = S^l, \quad VU = T^l \ . \ //$$

25. Corollary. If two topological Markov chains S, T are topologically conjugate then their defining matrices S, T are shift equivalent.

Is the converse of 25 valid i.e. does shift equivalence imply topological conjugacy? The answer to this (Williams's problem) is unknown. There is, however, another equivalent formulation of topological conjugacy called adapted shift equivalence, and perhaps some hope that shift equivalence may imply adapted shift equivalence:

If S is a $k \times k$ 0-1 matrix, we shall denote by S_n, $n \geq 0$, the 0-1 transition matrix for allowable S-words of length $n+1$: the rows and columns of S_n are indexed by the set

$$\{(i_0, i_1, \ldots, i_n) \in \{1, \ldots, k\}^{n+1} : S(i_0, i_1) = S(i_1, i_2) = \ldots = S(i_{n-1}, i_n) = 1\}$$

and $S_n(i_0, i_1, \ldots, i_n, j_0, j_1 \ldots j_n) = 1$ iff $j_0 = i_1$, $j_1 = i_2$, $\ldots$, $j_{n-1} = i_n$ and $S(i_n, j_n) = 1$. By definition, $S_0 = S$. It is easy to see that when S is irreducible and α is the state partition of the topological Markov chain defined by S we have $M(\alpha^n) = S_n$.

26. Definition. Two 0-1 matrices S, T are said to be adapted (or adapted shift equivalent) if there exists $n \in \mathbb{N}$ such that S_n and T_n are shift equivalent with lag n.

27. Theorem [P. 8]. Two topological Markov chains S, T are topologically conjugate iff S, T are adapted shift equivalent.

Proof. All the work for the proof in one direction has been done. Let α, α' be the state partitions of S, T respectively. If S and T are topologically conjugate then 19 and 24 show that, for some n, $M(\alpha^{2n}) = S_{2n}$ and $M(\alpha'^{2n}) = T_{2n}$ are shift equivalent with lag $2n$. In proving the converse we shall assume for convenience of presentation that the lag is 2: We assume

$$S_2 U = U T_2, \quad V S_2 = T_2 V, \quad UV = S_2^2, \quad VU = T_2^2$$

for some non-negative integral matrices U, V.

Consider the graphs associated with S_2 and T_2. U then specifies transitions from S_2-vertices to T_2-vertices and V specifies transitions from T_2-vertices to S_2-vertices. Arrows will indicate allowable transitions. S_2-vertices are triples, so that we have, for example

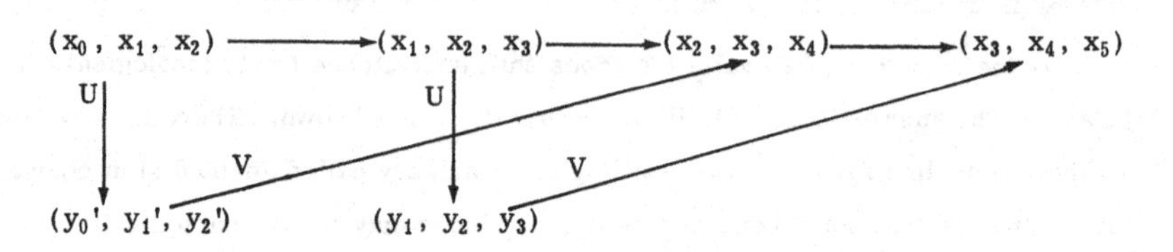

The question arises as to whether there is a transition from (y_0', y_1', y_2') to (y_1, y_2, y_3). By commutativity, there must exist y_0 such that

$$\begin{array}{l}(x_0, x_1, x_2) \\ \downarrow \\ (y_0, y_1, y_2) \longrightarrow (y_1, y_2, y_3)\end{array}$$

and therefore x such that

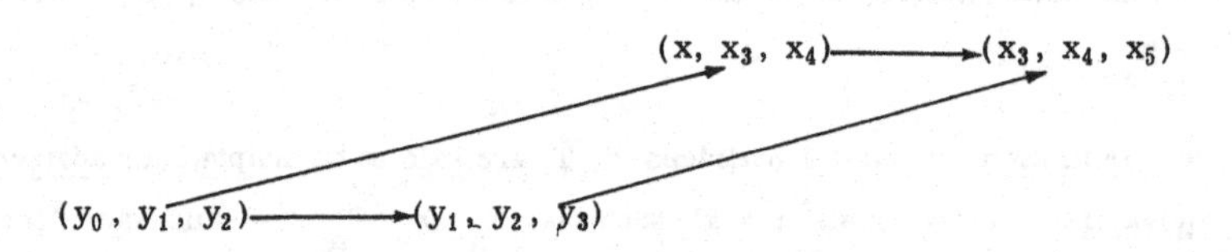

However x must necessarily be x_2 to comply with the commutativity

$$\begin{array}{l}(x_0, x_1, x_2) \longrightarrow (x_1, x_2, x_3) \longrightarrow (x_2, x_3, x_4) \\ \downarrow \nearrow \\ (y_0, y_1, y_2)\end{array}$$

Thus there are two paths from (x_0, x_1, x_2) to (x_2, x_3, x_4) passing through the T_2-vertices (y_0', y_1', y_2') and (y_0, y_1, y_2) which correspond to a unique S_2 path of length 2 $(x_0, x_1, x_2) \rightarrow (x_2, x_3, x_4)$. We conclude that $(y_0', y_1', y_2') \equiv (y_0, y_1, y_2)$ and that the transition $(y_0', y_1', y_2') \equiv (y_0, y_1, y_2) \rightarrow (y_1, y_2, y_3)$ is allowed.

For $x = (x_n)$ define $\varphi_0(x_0, x_1, x_2, x_3, x_4) = (y_0, y_1, y_2)$ where the transitions

$$\begin{array}{ccccc} (x_0, x_1, x_2) & \longrightarrow & (x_1, x_2, x_3) & \longrightarrow & (x_2, x_3, x_4) \\ \downarrow & & & \nearrow & \\ (y_0, y_1, y_2) & & & & \end{array}$$

are allowed. By the argument above,

$$(\phi(x)_n, \phi(x)_{n+1}, \phi(x)_{n+2}) = \varphi_0(x_n, x_{n+1}, x_{n+2}, x_{n+3}, x_{n+4})$$

then defines a continuous map ϕ such that $\phi S = T\phi$.

In a similar way we can define, for $y = (y_n)$, $\psi_0(y_0, y_1, y_2, y_3, y_4) = (x_2, x_3, x_4)$ where the transitions

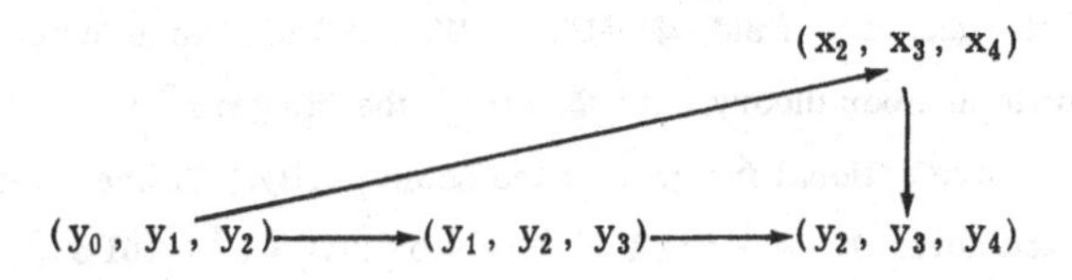

are allowed and, by $(\psi(y)_n, \psi(y)_{n+1}, \psi(y)_{n+2}) = \psi_0(y_n, y_{n+1}, y_{n+2}, y_{n+3}, y_{n+4})$, a continuous map ψ with $\psi T = S\psi$. It is clear that

$$\psi\phi = S^2, \quad \phi\psi = T^2.$$

Hence, ϕ and ψ are bijective, and are homeomorphisms. Thus S, T are topologically conjugate. //

4. INVARIANTS AND REVERSIBILITY

Having discussed algebraic formulations of topological conjugacy of topological Markov chains, we devote this section to invariants. As we shall see, none of our invariants clarifies Williams's problem - all computable invariants known to date (in particular topological entropy) are invariants of shift equivalence.

We start the section with an example of an irreversible topological Markov chain (one which is not topologically conjugate to its inverse). It will follow from this example that many invariants are not complete, since many invariants cannot distinguish between inverses. The example is due to Kōllmer. The proof presented here was prepared with the help of Les Davison. Cuntz and Krieger [C'. K'. 1] have since constructed many more examples.

Recall that the inverse of a topological Markov chain S is the topological Markov chain defined by the transpose matrix $S^{\star}$. We shall show that $S = \begin{pmatrix} 19 & 5 \\ 4 & 1 \end{pmatrix}$ and $S^{\star} = \begin{pmatrix} 19 & 4 \\ 5 & 1 \end{pmatrix}$ are not shift equivalent. In view of the results of last section and the exercise below, this is sufficient to conclude that the topological Markov chain S is not topologically conjugate to its inverse.

28. <u>Exercise.</u> Let S be a non-negative irreducible matrix. Let S_1 be the 0-1 transition matrix for paths in the graph obtained from S. Show that S and S_1 are (strong) shift equivalent with lag 1.

29. <u>Lemma.</u> <u>There are no non-negative integers</u> x, y <u>satisfying</u> $x^2 - 101y^2 = \pm 20$, $x - 9y \equiv 0 \pmod 4$, $x + 9y \equiv 0 \pmod 5$.

<u>Proof.</u> Consider the quadratic field $\mathbb{Q}(\sqrt{101})$. We shall use some standard terminology from algebraic number theory. In $\mathbb{Q}(\sqrt{101})$ the integers take the form $\frac{m}{2} + \frac{n}{2}\sqrt{101}$ where m, n are rational integers of the same parity. On the integers $\mathbb{Z}(\sqrt{101})$ of $\mathbb{Q}(\sqrt{101})$ the norm N is defined by $N(a + b\sqrt{101}) = a^2 - 101b^2$. N then satisfies $N((a+b\sqrt{101})(c+d\sqrt{101})) = N(a+b\sqrt{101})N(c+d\sqrt{101})$. $10 + \sqrt{101}$ is a unit of $\mathbb{Z}(\sqrt{101})$.

Suppose the non-negative rational integers x, y satisfy $N(x+y\sqrt{101}) = \pm 20$, $x - 9y \equiv 0 \pmod 4$, $x + 9y \equiv 0 \pmod 5$. Choose $n \geq 0$ so that

$$(10 + \sqrt{101})^n \leq x + \sqrt{101}\, y < (10 + \sqrt{101})^{n+1}$$

i.e.
$$1 \leq x_0 + \sqrt{101}\, y_0 < 10 + \sqrt{101}$$

where $x_0 + \sqrt{101}\, y_0 = (x + \sqrt{101}\, y)(-10 + \sqrt{101})^n$ and x_0, y_0 are rational integers. Since

$$\pm 20 = N(x + \sqrt{101}\, y) = (N(10 + \sqrt{101}))^n N(x_0 + \sqrt{101}\, y_0)$$
$$= (-1)^n N(x_0 + \sqrt{101}\, y_0)$$

we have $x_0^2 - 101y_0^2 = \pm 20$, or $\sqrt{101}\, y_0 = \pm \sqrt{(x_0^2 \pm 20)}$. Thus $1 \leq x_0 \pm \sqrt{(x_0^2 \pm 20)} < 10 + \sqrt{101}$. The possibilities are easily exhausted, we must have $x_0 = \pm 9$, $y_0 = 1$. Hence

$$x + \sqrt{101}\, y = (10 + \sqrt{101})^n (\pm 9 + \sqrt{101}).$$

Putting
$$x_{r+1} + \sqrt{101}\, y_{r+1} = (10 + \sqrt{101})(x_r + \sqrt{101}\, y_r)$$
$$= (10x_r + 101y_r) + \sqrt{101}\,(x_r + 10y_r),$$

we have $x = x_n$, $y = y_n$ with $(x_0, y_0) = (\pm 9, 1)$. Moreover

$$x_{r+1} = 10x_r + 101y_r,$$
$$y_{r+1} = x_r + 10y_r.$$

Now $x_{r+1} - 9y_{r+1} = x_r + 11y_r \equiv x_r - 9y_r \pmod{20}$ so that if $(x_0, y_0) = (-9, 1)$ then $x - 9y \equiv -18 \pmod{20}$, which contradicts $x - 9y \equiv 0 \pmod 4$. On the other hand

$$x_{r+1} + 9y_{r+1} = 19x_r + 191y_r \equiv 4x_r + 36y_r \equiv 4(x_r + 9y_r) \pmod 5$$

so that if $(x_0, y_0) = (9, 1)$ then $x + 9y \equiv 4^n . 3 \pmod 5$, contradicting $x + 9y \equiv 0 \pmod 5$. Therefore no non-negative integers satisfy $x^2 - 101y^2 = \pm 20$, $x - 9y \equiv 0 \pmod 4$, $x + 9y \equiv 0 \pmod 5$. //

30. Proposition. The matrix $S = \begin{pmatrix} 19 & 5 \\ 4 & 1 \end{pmatrix}$ is not shift equivalent to its transpose. Hence, the topological Markov chain defined by S is not topologically conjugate to its inverse.

Proof. Suppose S is shift equivalent to its transpose $S^\star$. Then there exist non-negative integral matrices U, V and $k \in \mathbb{N}$ such that $VU = S^k$ and $US = S^\star U$. Since $\det S = -1$, this means $\det U = \pm 1$. From $U\begin{pmatrix} 19 & 5 \\ 4 & 1 \end{pmatrix} = \begin{pmatrix} 19 & 4 \\ 5 & 1 \end{pmatrix} U$ we see that U must have the form $U = \begin{pmatrix} a & b \\ b & c \end{pmatrix}$ where $5a - 18b = 4c$. Thus $5a - 18b \equiv 0 \pmod 4$. As $\det U = \pm 1$ we also have $5a^2 - 18ab - 4b^2 = \pm 4$ i.e. $(5a - 9b)^2 - 101b^2 = \pm 20$. Now taking $x = 5a - 9b = 4c + 9b$ and $y = b$ we obtain $x^2 - 101y^2 = \pm 20$, $x - 9y \equiv 0 \pmod 4$, $x + 9y \equiv 0 \pmod 5$ for the non-negative integers x, y. According to 29 this is impossible. //

According to 4.8 topological entropy is invariant under bounded-to-one continuous conjugacies of topological Markov chains. Thus topological entropy is an invariant of topological conjugacy, but is far from complete.

Topological entropy and uniqueness of maximal measures, 2.26, enable us to use maximal measures as invariants of topological conjugacy: if the topological

Markov chains (X, S), (Y, T) are topologically conjugate by $\psi : X \to Y$ and if m_S and m_T denote their maximal measures, then $m_T = m_S \circ \psi^{-1}$. Hence the Markov chains (X, S, m_S) and (Y, T, m_T) are block isomorphic and the winding numbers group and homomorphism of Chapter IV can be used as invariants. 30 implies that these invariants are not complete since, by 4.18, the same invariants are attached to S and S^{-1}.

31. Exercise. Show that topological entropy and the winding numbers homomorphism are invariants of shift equivalence.

32. Exercise. Let S be the topological Markov chain defined by $\begin{pmatrix} 1 & 1 \\ 1 & 0 \end{pmatrix}$. Calculate its topological entropy and its maximal measure and the winding numbers group associated with this measure.

For each $n \geq 1$, the number of points of period n is clearly invariant under topological conjugacy i.e. the sequence of numbers $\{\theta_S(n)\}$, $\theta_S(n) = \text{card}\{x : S^n x = x\}$ is an invariant. The $\theta_S(n)$ can be incorporated in a function:

33. Definition. The zeta function of a topological Markov chain S is defined by

$$\zeta_S(t) = \exp\left(\sum_{n=1}^{\infty} \theta_S(n) t^n/n\right)$$

for $|t| < r$, where r is the radius of convergence of the power series.

34. Theorem [B. L.]. *Let S be an irreducible non-negative integral matrix with maximum eigenvalue β. Then the zeta function of the topological Markov chain S is*

$$\zeta_S(t) = 1/\det(I - tS), \quad |t| < \frac{1}{\beta}.$$

Proof. Note that $\theta_S(n) = \text{trace}(S^n)$. We have

$$\sum_{n=1}^{\infty} \theta_S(n) t^n/n = \sum_{n=1}^{\infty} t^n (\text{trace } S^n)/n$$
$$= \sum_{n=1}^{\infty} \frac{(t\lambda_1)^n + \dots + (t\lambda_k)^n}{n},$$

where $\lambda_1, \dots, \lambda_k$ are the eigenvalues of S. Thus for $|t| < \frac{1}{\beta}$,

$$\sum_{n=1}^{\infty} \theta_S(n) t^n/n = -\sum_{j=1}^{k} \log(1 - t\lambda_j)$$

$$= -\log \prod_{j=1}^{k} (1 - t\lambda_j) = -\log \det(I - tS)$$

and $\zeta_S(t) = 1/\det(I - tS)$. //

35. Corollary. If S, T are topologically conjugate topological Markov chains then $\det(I - tS) = \det(I - tT)$ for all $t \in \mathbb{R}$.

36. Exercise. Let S, T be irreducible non-negative integral matrices which are shift equivalent with lag l. Show that trace S^n = trace T^n for all $n \geq l$ and use 34 to deduce that $\det(I - tS) = \det(I - tT)$ for all $t \in \mathbb{R}$.

In effect, the characteristic polynomial of the matrix is the invariant, if we ignore multiples by a power of the indeterminate:

37. Corollary. If the irreducible non-negative integral matrices S, T are shift equivalent then their characteristic polynomials χ_S, χ_T satisfy $\chi_S(t) = t^n \chi_T(t)$ for some integer n. In particular $\chi_S(1) = \chi_T(1)$.

For instance, the matrices (2) and $\begin{pmatrix} 1 & 1 \\ 1 & 1 \end{pmatrix}$, which are (strong) shift equivalent, have characteristic polynomials $t - 2$ and $t(t - 2)$.

Again, 30 implies that the zeta function is not a complete invariant.

5. FLOW EQUIVALENCE

38. Definition. Let (X, S) be a topological Markov chain and let k be a strictly positive continuous function on X. Denote by X^k the compact metric space obtained from $\{(x, y) : x \in X, 0 \leq y \leq k(x)\}$ by identifying $(x, k(x))$ and $(Sx, 0)$ for each x. Let $\{S_t^k : t \in \mathbb{R}\}$ be the flow on X^k obtained by moving each (x, y) vertically at unit velocity until $(x, k(x)) \sim (Sx, 0)$ is reached whereupon the flow resumes vertically from $(Sx, 0)$. (X^k, S_t^k) is called the k-suspension of (X, S). When $k \equiv 1$, (X^1, S_t^1) is called the standard suspension, or simply the suspension, of (X, S).

In 38 the region under the function $k(x)$ is a fundamental region for an equivalence relation on $X \times \mathbb{R}$: For positive integers n define $k(x, n) = k(x) + k(Sx) + \ldots + k(S^{n-1}x)$ and take $k(x, 0) = 0$. Then

$$k(x,\ m+n) = k(x,\ n) + k(S^n x,\ m) \qquad (\bigstar)$$

Define $k(x, -n) = -k(S^{-n}x, n)$ so that $(\bigstar)$ holds for all integers m, n. For $(x_1, y_1), (x_2, y_2) \in X \times \mathbb{R}$ put $(x_1, y_1) \sim (x_2, y_2)$ iff $x_2 = S^n x_1$ and $y_1 - y_2 = k(x_1, n)$ for some integer n. That $\sim$ is an equivalence relation follows from the definition of $k(x, n)$ and $(\bigstar)$. It is now easy to see that the region under k is a fundamental region for $\sim$ so that S_t^k may be regarded as the flow induced on $X \times \mathbb{R}/\sim$ by the vertical flow $(x, y) \mapsto (x, y+t)$ on $X \times \mathbb{R}$.

Flow equivalence of topological Markov chains will be defined in terms of standard suspensions. (Compare 38 with 4.12.)

Recall that two flows $\{S_t\}$ and $\{T_t\}$ are said to be topologically conjugate if there exists a homeomorphism ϕ such that $\phi S_t = T_t \phi$ for all $t \in \mathbb{R}$.

39. Proposition. Let (X, S) be a topological Markov chain. If $k, h : X \to \mathbb{R}$ are positive functions which are cohomologous (i.e. $k = h + g \circ S - g$ for some continuous g) then $\{S_t^k\}$ and $\{S_t^h\}$ are topologically conjugate flows.

Proof. Associated with k, h we have equivalence relations $\sim_k, \sim_h$ on $X \times \mathbb{R}$. Consider the homeomorphism ϕ of $X \times \mathbb{R}$ defined by $(x, y) \mapsto (x, y+g(x))$. ϕ respects the equivalence relations $\sim_k, \sim_h$, for if $x_2 = S^n x_1$ and $y_1 - y_2 = k(x_1, n)$ then $y_1 + g(x_1) - y_2 - g(x_2) = k(x_1, n) + g(x_1) - g(S^n x_1) = h(x_1, n)$. Hence, ϕ induces a homeomorphism of $X \times \mathbb{R}/\sim_k$ onto $X \times \mathbb{R}/\sim_h$. It is easy to check that this homeomorphism conjugates the flows $\{S_t^k\}$ and $\{S_t^h\}$. //

If (X, S) is a topological Markov chain with state partition $\mathfrak{a}$, we shall say that a function $h : X \to \mathbb{R}$ depends only on a finite number of past coordinates if h is measurable with respect to $\bigvee_{i=0}^{n} S^{-i}\mathfrak{a}$ for some $n \in \mathbb{N}$.

40. Theorem [P. 4]. Let (X, S) be a topological Markov chain and let k be a positive continuous function. Suppose $\{S_t^k\}$ has a continuous eigenfunction f with eigenfrequency $a > 0$ (i.e. $f \circ S_t^k = e^{2\pi i a t} f$ for all $t \in \mathbb{R}$). Then k is cohomologous to

(i) a continuous function which depends only on a finite number of past coordinates and which assumes only integral multiples of a^{-1} for values,

(ii) a positive continuous function which depends only on a finite number of past coordinates and which assumes only integral multiples of $(na)^{-1}$ for values, where n is a fixed positive integer.

Proof. The topologically transitive homeomorphism S has a dense orbit $\{S^n x_0 : n \in \mathbb{Z}\}$. Now the orbit $\{S_t^k(x_0, 0) : t \in \mathbb{R}\}$ is dense in X^k and, since $|f| \circ S_t^k = |f|$, $|f|$ is constant on this dense orbit. Hence $|f|$ is constant, and we may assume $|f| = 1$. Now by 4.14 f is homotopic to a function of the form

$$(x, y) \mapsto \exp 2\pi i \frac{M(x)}{k(x)} y \quad (0 \leq y \leq k(x))$$

where $M : X \to \mathbb{Z}$ is continuous. $f(x, y) \exp(-2\pi i \frac{M(x)}{k(x)} y)$ is thus homotopic to 1 and we may write $f(x, y) = \exp 2\pi i(\frac{M(x)}{k(x)} y + l(x, y))$ for some continuous function $l : X^k \to \mathbb{R}$. Since $\frac{1}{2\pi i} \frac{f'}{f} = a$, where f' denotes the derivative of $f \circ S_t^k$ with respect to t at 0, we see that

$$\frac{M(x)}{k(x)} + l'(x, y) = a$$

where the derivative l' is with respect to the second variable. Hence $l(x, y) = (a - \frac{M(x)}{k(x)}) y + \theta(x)$ for some continuous $\theta : X \to \mathbb{R}$. Since $l(x, k(x)) = l(Sx, 0)$ we conclude that

$$k(x) = \frac{M(x)}{a} + \frac{\theta(Sx)}{a} - \frac{\theta(x)}{a} .$$

The integer valued function M on the compact space X can assume only a finite number of values, and for each of these values n, $M^{-1}(n)$ is a closed-open set. Hence each $M^{-1}(n)$ is a finite union of cylinders and M is measurable with respect to $\bigvee_{i=-p}^{p} S^i \alpha$ for some $p \in \mathbb{N}$, where α is the state partition. Put $N = M \circ S^p$. The function $\frac{N}{a}$ is cohomologous to k and satisfies all the requirements of (i). For (ii), write $N(x) = ak(x) + \theta_1(Sx) - \theta_1(x)$ where $\theta_1 : X \to \mathbb{R}$ is continuous. Since also $a > 0$ and k is continuous and strictly positive we have for some positive integer n,

$$N(x, n) = ak(x, n) + \theta_1(S^n x) - \theta_1(x) > 0 \text{ for all } x \in X.$$

However, $k(x, n)$ is cohomologous to $nk(x)$ and $\theta_1(S^n x) - \theta_1(x)$ is a coboundary. Consequently $N(x, n)$ is cohomologous to $nak(x)$ i.e.

$$k(x) = \frac{1}{na}(N(x, n)) + \theta_2(Sx) - \theta_2(x)$$

for some continuous function θ_2, and $N(X, n) = N(X) + N(Sx) + \ldots + N(S^{n-1}x)$ satisfies the requirements of (ii). //

41. Definition. If $\{S_t\}$, $\{T_t\}$ are flows on the compact spaces X, Y respectively, then $\{S_t\}$, $\{T_t\}$ are said to be flow equivalent if there exists a homeomorphism $\phi : X \to Y$ which sends orbits of S_t to orbits of T_t, preserving orientation but not necessarily parametrisation. Two topological Markov chains S, T are said to be flow equivalent if their standard suspensions are flow equivalent.

The following proposition is easy to prove:

42. Proposition. In order that two topological Markov chains (X, S) and (Y, T) be flow equivalent it is necessary and sufficient that there is a strictly positive continuous function $h : X \to \mathbb{R}$ such that $\{S_t^h\}$ and $\{T_t^1\}$ are topologically conjugate.

43. Theorem [P. 4]. If S, T are flow equivalent topological Markov chains then for some positive continuous rational valued function k depending only on a finite number of past coordinates $\{S_t^k\}$ and $\{T_t^1\}$ are topologically conjugate flows.

Proof. By 42, there is a positive continuous function h such that $\{S_t^h\}$ and $\{T_t^1\}$ are topologically conjugate. Since $\{T_t^1\}$ has an eigenfunction with eigenfrequency 1, the same is true of $\{S_t^h\}$. Now 40 provides a positive continuous rational valued function k which depends only on a finite number of past coordinates and which is cohomologous to h. Finally 39 shows that $\{S_t^h\}$ and $\{S_t^k\}$ are topologically conjugate. //

Suppose S, T are flow equivalent topological Markov chains. By replacing S by a suitable one of the strong shift equivalent matrices S_n $(n \in \mathbb{N})$ defined in Section 3, we assume that the function k in 43 depends on only one coordinate i.e. $k(x) = k(x_0)$ for $x = (x_n)$. Let N be a (positive) common denominator of the rational values of k. Thus, if S is $l \times l$, for each $i \in \{1, \ldots, l\}$ we can write $k(i) = \frac{n(i)}{N}$ with $n(i) \in \mathbb{Z}$, $n(i) > 0$. We use the n(i) to define a new irreducible 0-1 transition matrix S' : S' has $\sum_{i=1}^{l} n(i)$ vertices divided into l groups with n(i) vertices in group i. In group i each vertex (except the top) leads only to the vertex above it. The top vertex leads to the bottom vertex of group j iff $S(i, j) = 1$. From T we obtain an irreducible 0-1 transition matrix T': If T is $m \times m$ then T' has Nm vertices divided into m groups of N vertices. The top vertex in group i leads to the bottom vertex of group j whenever $T(i, j) = 1$. All other vertices lead only to the vertex above them.

Consider now the homeomorphism $S^k_{1/N}$. Clearly, $S^k_{1/N}$ can be regarded as the direct product of the topological Markov chain S' with the identity on the interval $[0, \frac{1}{N}]$. Similarly $T^1_{1/N}$ can be regarded as the direct product of the topological Markov chain T' with the identity on $[0, \frac{1}{N}]$. Moreover, by 43, $T^1_{1/N}$ and $S^k_{1/N}$ are topologically conjugate. It follows that S', T' are topologically conjugate topological Markov chains. Combining this with 20 we arrive at

44. Theorem [P. S']. *Two topological Markov chains* S, T *are flow equivalent iff it is possible to get from the matrix* S *to the matrix* T *by a finite sequence of the following operations on non-negative integral (square) matrices:*

(i) *Replace a (product) matrix* UV *by* VU, *where* U, V *are rectangular non-negative integral matrices.*

(ii) *Replace* $M = \begin{pmatrix} M(1,1) & \dots & M(1,n) \\ \vdots & & \vdots \\ M(n,1) & \dots & M(n,n) \end{pmatrix}$ *by*

$$M' = \begin{pmatrix} 0 & M(1,1) & \dots & M(1,n) \\ 1 & 0 & \dots & 0 \\ 0 & M(2,1) & \dots & M(2,n) \\ & \vdots & & \\ 0 & M(n,1) & \dots & M(n,n) \end{pmatrix}$$

(iii) *The inverse of operation (ii).*

44 enables us to establish as an invariant the value at 1 of the characteristic polynomial of the defining matrix:

45. Corollary [P. S'.]. *If* S, T *are flow equivalent topological Markov chains then* $\chi_S(1) = \chi_T(1)$.

Proof. In view of 37, we need only prove that the matrices M, M' in operation (ii) satisfy $\chi_M(1) = \chi_{M'}(1)$. But

$$\chi_{M'}(1) = \det(I - M') = \det \begin{pmatrix} 1 & -M(1, 1) & -M(1, 2) & \dots & -M(1, n) \\ -1 & 1 & 0 & \dots & 0 \\ 0 & -M(2, 1) & 1-M(2, 2) & \dots & -M(2, n) \\ & \vdots & & & \\ 0 & -M(n, 1) & -M(n, 2) & \dots & 1-M(n, n) \end{pmatrix}$$

$$= \det \begin{pmatrix} 1 & 1-M(1, 1) & -M(1, 2) & \dots & -M(1, n) \\ -1 & 0 & 0 & & 0 \\ 0 & -M(2, 1) & 1-M(2, 2) & \dots & -M(2, n) \\ & \vdots & & & \\ 0 & -M(n, 1) & -M(n, 2) & \dots & 1-M(n, n) \end{pmatrix}$$

and, expanding by the first column,

$$\chi_{M'}(1) = \det(I - M) = \chi_M(1). \quad //$$

That many topological Markov chains cannot be flow equivalent follows from 45. For instance, if $m \neq n$ then the full m-shift and the full n-shift are not flow equivalent.

Let S be an $n \times n$ integral matrix and let A be an Abelian group. S acts as a homomorphism of A^n into itself by

$$S(a) = S \begin{pmatrix} a_1 \\ \vdots \\ a_n \end{pmatrix} = \begin{pmatrix} \sum_{j=1}^{n} S(1, j) a_j \\ \vdots \\ \sum_{j=1}^{n} S(n, j) a_j \end{pmatrix}.$$

Put $\mathrm{Fix}_S(A) = \{a \in A^n : S(a) = a\}$. The following is a direct consequence of 44:

46. Corollary [B. F.]. *If S, T are flow equivalent topological Markov chains then for each Abelian group A, $\mathrm{Fix}_S(A)$ and $\mathrm{Fix}_T(A)$ are isomorphic groups.*

47. Exercise. Let $S^{\star}$ denote the transpose of the integral matrix S. Let A be an Abelian group. Show that the groups $\mathrm{Fix}_S(A)$ and $\mathrm{Fix}_{S^{\star}}(A)$ are isomorphic.

The invariants of flow equivalence provided by 45 and 46 fail to distinguish between topological Markov chains and their inverses. Thus the following problem remains. Are topological Markov chains flow equivalent to their inverses? A possible (though unlikely) way of answering this question is to extend 46 to Abelian semi-groups under the assumption that S is non-negative and irreducible. That irreducibility is necessary can be seen from the following example:

Let $S = \begin{pmatrix} 1 & 1 \\ 0 & 2 \end{pmatrix}$ and let A be the Abelian semi-group of non-negative integers under addition. Then

$$\mathrm{Fix}_S(A) = \left\{ \begin{pmatrix} a \\ b \end{pmatrix} : \begin{pmatrix} a \\ b \end{pmatrix} = \begin{pmatrix} 1 & 1 \\ 0 & 2 \end{pmatrix} \begin{pmatrix} a \\ b \end{pmatrix} = \begin{pmatrix} a+b \\ 2b \end{pmatrix} \right\} = \left\{ \begin{pmatrix} a \\ 0 \end{pmatrix} : a \in A \right\}$$

whereas

$$\mathrm{Fix}_{S^*}(A) = \left\{ \begin{pmatrix} a \\ b \end{pmatrix} : \begin{pmatrix} a \\ b \end{pmatrix} = \begin{pmatrix} 1 & 0 \\ 1 & 2 \end{pmatrix} \begin{pmatrix} a \\ b \end{pmatrix} = \begin{pmatrix} a \\ a+2b \end{pmatrix} \right\} = \left\{ \begin{pmatrix} 0 \\ 0 \end{pmatrix} \right\}.$$

48. Remarks. (i) The work on subshifts of finite type was pioneered by Hedlund's work on full shifts (see [H']). In addition to the papers mentioned in this chapter, [C. P.] and [M] contain important work on subshifts of finite type. For the relationship between subshifts of finite type and Axiom A diffeomorphisms, see [B. 1]. Krieger, Cuntz and others have been pursuing the connection between topological Markov chains and C^*-algebras (see, for instance, [K'], [C'. K'. 1], [C'. K'. 2], [C'. E.]).

(ii) Most of the theory in Sections 1, 3, 4 and 5 can be paralleled for Markov chains and block-codes (see [P. T. 1] and [P. T. 2]). It is conjectured in [P. T. 1] that the pressure invariants of 2.31 and 4.11 are complete invariants for finite equivalence of Markov chains. A classification, by these pressure invariants and period, analogous to 12 is, however, not possible (see the end of Section 5, Chapter II). Kitchens [K] has recently obtained interesting results for zeta functions.

APPENDIX: SHANNON'S WORK ON MAXIMAL MEASURES

We repeat 2.26 as

1. Theorem [P. 2]. Let (X, T) be a topological Markov chain. There is a unique T-invariant probability m such that $h_m(T) \geq h_\mu(T)$ for all T-invariant Borel probabilities μ. m is Markov and is supported by X.

This result was proved in [P. 2] without the knowledge that Shannon had included a similar theorem in his 1948 paper [S. W.]. In this Appendix we interpret the relevant parts of Shannon's paper to compare his theorem with 1. We will show that Shannon proved:

2. Theorem. Let (X, T) be an aperiodic topological Markov chain. There is a Markov probability m on X such that $h_m(T) \geq h_\mu(T)$ for all (compatible) Markov probabilities μ on X.

Comparing 2 with 1, we notice that in 2 μ is allowed to run through only Markov probabilities. This (insignificant) restriction is natural since at the time Shannon wrote, entropy had not been defined in general - in [S. W.] he defined it, for the first time, in some special cases. More significantly, there is no explicit uniqueness statement in 2.

All of Shannon's work is in the setting of (a model of) a communication system and in fact he proves 2 for systems (superficially) more general than aperiodic topological Markov chains. We consider Shannon's communication system and state his theorem for such a system. Then we show that this theorem is equivalent to 2, and prove it by Shannon's method.

Suppose we have a collection $\mathcal{S} = \{s_1, \ldots, s_n\}$ of "symbols" and a collection $\{a_1, \ldots, a_m\}$ $(1 \leq m \leq n)$ of "states". With each a_i $(1 \leq i \leq m)$ associate a set $\mathcal{S}(a_i) \subset \mathcal{S}$, of symbols that may be transmitted when we are in the state a_i. With each $s_j \in \mathcal{S}$ associate a state $a(s_j) \in \{a_1, \ldots, a_m\}$. Think of $a(s_j)$ as the state we are in after a transmission of s_j. Each s_j also has a number $l_j \in \mathbf{N}$, $l_j \geq 1$, associated with it. l_j is the "length of s_j" or the "time taken to transmit s_j". A <u>(finite) signal</u> is a (finite) sequence (s_{i_k}) of symbols such that $s_{i_{k+1}} \in \mathcal{S}(a(s_{i_k}))$. Strictly speaking, a finite signal is a starting state a_i followed by a finite sequence of symbols $s_{i_1}, s_{i_2}, \ldots, s_{i_l}$ such that $s_{i_1} \in \mathcal{S}(a_i)$ and, for $2 \leq k \leq l$, $s_{i_k} \in \mathcal{S}(a(s_{i_{k-1}}))$. The space we are working with, though not explicitly stated in [S. W.], is that of all signals with the shift and the topology generated by the finite signals.

The system described in the above paragraph is a <u>discrete (noiseless) communication system.</u> This is the system Shannon defines and works with. His definition is motivated by telegraphy:

In telegraphy there are, at "machine level", two distinct symbols - the line open (for unit time) and the line closed (for unit time). For convenience we overlook this and assume that (at a higher level) the symbols are:

(i) dot, consisting of line closure for unit time followed by line open for unit time,

(ii) dash, consisting of three units of line closure followed by one unit open,

(iii) letter space, line open for three units,

(iv) word space, line open for six units.

The corresponding times of transmission are 2, 4, 3, 6. The only constraint is that spaces may not follow each other. Thus, we have two states a_1 and a_2; $\mathcal{S}(a_1)$ consists of all symbols and $\mathcal{S}(a_2) = \{\text{dot, dash}\}$. $a(\text{dot}) = a(\text{dash}) = a_1$ (i.e. we are in state a_1 after the transmission of a dot or a dash) and $a_2 = a(\text{letter space}) = a(\text{word space})$. This may be represented graphically as:

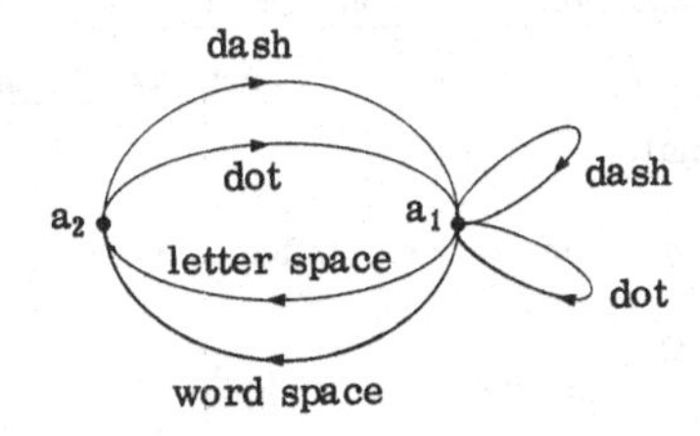

Take a discrete communication system with symbols $\mathcal{S}$ and states $\{a_1, \ldots, a_m\}$. Accessibility of a state from another is defined in the obvious way. We obtain an $m \times m$ 0-1 matrix A by putting $A_{ij} = 1$ iff a_j is accessible from

a_i in one step. We refer to A as the matrix of the system. The system is called irreducible (resp. aperiodic) if its matrix is irreducible (resp. aperiodic). From now on we only consider aperiodic systems. This corresponds to the assumptions made by Shannon on p. 17-18 of [S. W.]. On aperiodic systems we shall consider measures obtained from a set of transition probabilities $P_{ij}^{(s)}$, $s \in \mathcal{S}, 1 \leq i, j \leq m$. $P_{ij}^{(s)}$ is the probability of passing from the state a_i to a_j (in one step) by transmitting the symbol s. $P_{ij} = \sum_{s \in \mathcal{S}} P_{ij}^{(s)}$ is thus the probability of the state a_j following a_i. Of course, we assume that the probabilities are compatible with the system i.e. that $P_{ij}^{(s)} > 0$ iff $s \in \mathcal{S}(a_i)$ and $a(s) = a_j$. It follows that $P_{ij} > 0$ iff $A_{ij} = 1$. By irreducibility, the $m \times m$ stochastic matrix P has an invariant probability vector p, $pP = p$. We obtain an invariant Borel probability μ on the system by defining, for a finite signal $F = s_1 \dots s_k$ starting at the state a_i,

$$\mu(F) = p_i P_{ii_1}^{(s_1)} P_{i_1 i_2}^{(s_2)} \dots P_{i_{k-1} i_k}^{(s_k)} ,$$

where $a_{i_j} = a(s_j)$, $1 \leq j \leq k$. $l^{(s)}$ will denote the length of the symbol s.

For an (aperiodic) discrete communication system with symbols $\mathcal{S}$, states $\{a_1, \dots, a_n\}$ and an invariant probability μ given by the compatible $\{P_{ij}^{(s)}\}$, the entropy (per unit time) is defined to be

$$H_\mu = \frac{-\sum_{i,j,s} p_i P_{ij}^{(s)} \log P_{ij}^{(s)}}{\sum_{i,j,s} p_i P_{ij}^{(s)} l^{(s)}} .$$

Shannon's theorem is ([S. W.], Appendix 4) :

3. Theorem (Shannon). For an aperiodic discrete communication system there is a choice of compatible $\{P_{ij}^{(s)}\}$ such that for the measure μ given by this choice H_μ is maximal.

For a system in which all symbols have length 1,

$$\sum_{i,j,s} p_i P_{ij}^{(s)} l^{(s)} = \sum_{i,j,s} p_i P_{ij}^{(s)} = \sum_{i,j} p_i P_{ij} = 1$$

for any compatible $\{P_{ij}^{(s)}\}$. Thus in this case $H_\mu = -\sum_{i,j,s} p_i P_{ij}^{(s)} \log P_{ij}^{(s)}$ which is (in today's terms) the entropy of the shift with respect to the "state partition", the partition given by all possible pairs (a_i, s) with $s \in \mathcal{S}(a_i)$. Given any discrete communication system we may, by introducing new states along its symbols of length greater than one, obtain an equivalent system in which all symbols have length 1. In other words we may, and shall, restrict our attention to systems in which all symbols have length 1. The following paragraph discusses the connection between such systems and topological Markov chains; it should be clear from this that 2 and 3 are equivalent.

If (X, T) is a topological Markov chain defined by the $n \times n$ irreducible 0-1 matrix A, then (X, T) may be regarded as a discrete communication system by taking $\{1, \ldots, n\}$ as symbols all of which have length 1 and a state for each symbol. So there is no real distinction between states and symbols and H_μ defined above reduces to the familiar $H_\mu = -\sum_{i,j} p_i P_{ij} \log P_{ij}$ for each Markov measure. It may happen that some rows of A are identical, and we may wish to consider only those Markov measures that are given by (compatible) matrices which preserve this property. Discrete communication systems are precisely the structures that cater for this; if $\{i_1, \ldots, i_k\} \subset \{1, \ldots, n\}$ are the (indices of) identical rows of A which we want to remain identical in any compatible stochastic matrix we consider, we combine the associated symbols into a common state i.e. take $a(i_1) = \ldots = a(i_k)$. The notation $P_{ij}^{(s)}$ is suited to this. Thus, discrete communication systems in which all symbols have length 1 are just topological Markov chains viewed in a way suitable for considering certain types of Markov measures.

For a finite signal in a discrete communication system, the *length* (or "time of transmission") of the signal is the sum of the lengths of the symbols in the signal. The *capacity* of the system is defined by

$$C = \lim_{t \to 0} \frac{\log N(t)}{t}$$

where $N(t)$ is the number of finite signals of length t. When all signals have length 1, this is just the topological entropy. We now proceed to Shannon's proof of 3 in this case.

Lemma. *Let* A *be an aperiodic non-negative* $n \times n$ *matrix and let* $\beta > 0$ *be its maximum eigenvalue. If* $\{x_k\}$ *is a positive sequence in* $\mathbb{R}^n$ *satisfying* $x_{k+1} = Ax_k$, *then there is an eigenvector* $r = (r(1), \ldots, r(n))^{tr}$ *corresponding to* β *such that* $\lim_{k \to \infty} \frac{x_k(i)}{r(i)\beta^k} = 1$ *for* $i = 1, \ldots, n$.

Proof. Write $B = M^{-1}AM$ where B is the Jordan canonical form of A. Since β is a simple eigenvalue, we may assume $B = \begin{pmatrix} \beta & & & \\ & J_1 & & 0 \\ & & \ddots & \\ & 0 & & J_m \end{pmatrix}$ where each

$J_i = \begin{pmatrix} \alpha & 1 & & & 0 \\ & \alpha & 1 & & \\ & & \ddots & \ddots & \\ & & & \alpha & 1 \\ 0 & & & & \alpha \end{pmatrix}$ for some eigenvalue $\alpha = \alpha_i \neq \beta$. Since A is aperiodic, this means $|\alpha| < \beta$. Put $y_k = M^{-1}x_k$. Then $y_{k+1} = By_k$. Writing $y_k = (y_k^{(0)}, y_k^{(1)}, \ldots, y_k^{(m)})^{tr}$ where the length of $y_k^{(0)}$ is 1 and the length of each $y_k^{(i)}$ $(1 \leq i \leq m)$ is the column (row) length of J_i, we see that $y_{k+1}^{(i)} = J_i y_k^{(i)}$. If J_i is $r \times r$ and corresponds to the eigenvalue $\alpha \neq 0$, it can be checked that $y_k^{(i)}$ must be a linear combination of the columns of the $r \times r$ matrix

$$\begin{pmatrix} \alpha^k & \binom{k}{1}\alpha^{k-1} & \binom{k}{2}\alpha^{k-2} & \ldots & \binom{k}{r-1}\alpha^{k-r+1} \\ 0 & \alpha^k & \binom{k}{1}\alpha^{k-1} & \ldots & \binom{k}{r-2}\alpha^{k-r+2} \\ 0 & 0 & \alpha^k & & \binom{k}{r-3}\alpha^{k-r+3} \\ \vdots & \vdots & \vdots & & \vdots \\ 0 & 0 & 0 & & \alpha^k \end{pmatrix}.$$

If J_i corresponds to $\alpha = 0$, $y_k^{(i)}$ is eventually the zero vector. Since $|\alpha| < \beta$, now writing $y_k = (y_k(1), \ldots, y_k(n))^{tr}$, we have $\lim_{k\to\infty} (y_k(i)/\beta^k) = 0$ for $2 \leq i \leq n$. Hence

$$\lim_{k\to\infty} (x_k/\beta^k) = \lim_{k\to\infty} (My_k/\beta^k) = M(c, 0, \ldots, 0)^{tr}$$

for some non-zero constant c and the result follows from this. //

Lemma ([S. W.], Appendix 1). For an aperiodic discrete communication system (whose symbols all have length 1) with matrix A, the capacity is $C = \log \beta$ where $\beta > 0$ is the maximum eigenvalue of A.

Proof. Let $\{a_1, \ldots, a_n\}$ be the states of the system. Let $N_i(L)$ be the number of finite signals of length L that start at the state a_i. We put $N(L) = (N_1(L), \ldots, N_n(L))^{tr}$. Since all symbols have length one, $N(L+1) = AN(L)$. By the preceding lemma we can find an eigenvector $(r(1), \ldots, r(n))^{tr}$ corresponding to β such that, for $1 \leq i \leq n$ we have $\frac{N_i(L)}{r(i)\beta^L} \to 1$ as $L \to \infty$. Thus,

$$C = \lim_{L\to\infty} \frac{1}{L} \log\Big(\sum_{i=1}^{n} N_i(L)\Big) = \lim_{L\to\infty} \frac{1}{L} \log\Big(\sum_{i=1}^{n} r(i)\beta^L\Big) = \log \beta . \quad /\!/$$

If μ is the measure given by the compatible set of transition probabilities $\{p_{ij}^{(s)}\}$, put $G_k = -\frac{1}{k}\sum_B \mu(B) \log \mu(B)$ where the sum is over all finite signals B of length k. Then $G_k \to H_\mu$ as $k \to \infty$. Now simply note that $G_k \leq \frac{\log N(k)}{k}$, where $N(k)$ is the number of finite signals of length k, to obtain $H_\mu \leq C$. This establishes C as an upper bound for the entropies.

If the $n \times n$ matrix A is the matrix of the system and $\beta > 0$ is the maximum eigenvalue of A with corresponding strictly positive eigenvector $r = (r(1), \ldots, r(n))^{tr}$, define the compatible probabilities $P_{ij}^{(s)}$ by

$$P_{ij}^{(s)} = \begin{cases} 0 & \text{if } s \notin S(a_i) \text{ or } a_j \neq a(s) \\ \dfrac{r(j)}{\beta r(i)} & \text{otherwise}. \end{cases}$$

Then $-\sum_{i,j,s} p_i P_{ij}^{(s)} \log P_{ij}^{(s)} = \log \beta = C$ and this completes the proof of 3 by Shannon's method.

REFERENCES

[A. G. W.] R. L. Adler, L. W. Goodwyn and B. Weiss, Equivalence of topological Markov shifts, Israel J. Math., 27 (1977), 49-63.

[A. M.] R. L. Adler and B. Marcus, Topological entropy and equivalence of dynamical systems, Mem. A. M. S., 219 (1979).

[A. W.] R. L. Adler and B. Weiss, Similarity of automorphisms of the torus, Mem. A. M. S., 98 (1970).

[A. J. R.] M. A. Akçoğlu, A. del Junco and M. Rahe, Finitary codes between Markov processes, Z. Wahrscheinlichkeitstheorie, 47 (1979), 305-14.

[B] P. Billingsley, Ergodic Theory and Information, Wiley, N. Y. (1965).

[B. 1] R. Bowen, Equilibrium States and the Ergodic Theory of Anosov Diffeomorphisms, S. L. N. 470, Springer, N. Y. (1975).

[B. 2] R. Bowen, Smooth partitions of Anosov diffeomorphisms are weak Bernoulli, Israel J. Math., 21 (1975), 95-100.

[B. F.] R. Bowen and J. Franks, Homology for zero-dimensional non-wandering sets, Ann. of Math., 106 (1977), 73-92.

[B. L.] R. Bowen and O. E. Lanford, Zeta functions of restrictions of the shift transformation, Proc. Symp. Pure Math., A. M. S. 14 (1970), 43-50.

[B. R.] R. Bowen and D. Ruelle, The ergodic theory of Axiom A flows, Invent. Math., 29 (1975), 181-202.

[C. P.] E. M. Coven and M. E. Paul, Endomorphisms of irreducible subshifts of finite type, Math. Syst. Theory, 8 (1974), 167-75.

[C'. E.] J. Cuntz and D. E. Evans, Some remarks on the C^*-algebras associated with certain topological Markov chains, to appear in Math. Scand.

[C'. K'. 1] J. Cuntz and W. Krieger, Topological Markov chains with dicyclic dimension groups, J. für die Reine und Angew. Math., 320 (1980), 44-51.

[C'. K'. 2] J. Cuntz and W. Krieger, A class of C^*-algebras and topological Markov chains, Invent. Math., 56 (1980), 251-68.

[F] W. Feller, An Introduction to Probability Theory and its Applications, Vol. 1, Wiley, N.Y. (1970).

[F.P.] R. Fellgett and W. Parry, Endomorphisms of a Lebesgue space II, Bull. L.M.S., 7 (1975), 151-8.

[F.O.] N. Friedman and D. S. Ornstein, On the isomorphism of weak Bernoulli transformations, Adv. in Math., 5 (1970), 365-94.

[H] P. R. Halmos, Lectures in Ergodic Theory, Chelsea, N.Y. (1958).

[H'] G. A. Hedlund, Endomorphisms and automorphisms of the shift dynamical system, Math. Syst. Theory, 3 (1969), 320-75.

[J] A. del Junco, Finitary codes between one-sided Bernoulli shifts, to appear in Ergodic Theory and Dynamical Syst.

[K.S.1] M. Keane and M. Smorodinsky, A class of finitary codes, Israel J. Math., 26 (1977), 352-71.

[K.S.2] M. Keane and M. Smorodinsky, Bernoulli schemes of the same entropy are finitarily isomorphic, Ann. of Math., 109 (1979), 397-406.

[K.S.3] M. Keane and M. Smorodinsky, Finitary isomorphism of irreducible Markov shifts, Israel J. Math., 34 (1979), 281-6.

[K] B. Kitchens, Proc. A.M.S., 83 (1981), 825-8.

[K'] W. Krieger, On dimension functions and topological Markov chains, Invent. Math., 56 (1980), 239-50.

[K.M.T.] I. Kubo, H. Murata and H. Totoki, On the isomorphism problem for endomorphisms of Lebesgue spaces I, II, III, Publ. Res. Inst. Math. Sci., 9 (1973/74), 285-317.

[L.R.] O. E. Lanford and D. Ruelle, Observables at infinite and states with short range correlations in statistical mechanics, Comm. Math. Phys., 13 (1969), 194-215.

[M] B. Marcus, Factors and extensions of full shifts, Monats. Math., 88 (1979), 239-47.

[M'] L. Meschalkin, A case of isomorphism of Bernoulli schemes, Dokl. Akad. Nauk. S.S.S.R., 128 (1959), 41-4 and Soviet Math. Dokl., 3 (1962), 1725-9.

[O.1] D. S. Ornstein, Ergodic Theory, Randomness and Dynamical Systems, Yale Univ. Press, New Haven (1974).

[O.2] D. S. Ornstein, Bernoulli shifts with the same entropy are isomorphic, Adv. in Math., 4 (1970), 337-52.

[P.P.W.] R. Palmer, W. Parry and P. Walters, Large sets of endomorphisms and of g-measures, Proc. North Dakota State Univ., June 1977, S.L.N. 668, Springer, N.Y. (1978).

[P.1] W. Parry, Topics in Ergodic Theory, Cambridge Tracts in Math. 75, C.U.P., Cambridge (1981).

[P.2] W. Parry, Intrinsic Markov chains, Trans. A.M.S., 112 (1964), 55-66.

[P.3] W. Parry, Endomorphisms of a Lebesgue space III, Israel J. Math., 21 (1975), 167-72.

[P.4] W. Parry, Topological Markov chains and suspensions, Univ. of Warwick notes.

[P.5] W. Parry, A finitary classification of topological Markov chains and sofic systems, Bull. L.M.S., 9 (1977), 86-92.

[P.6] W. Parry, The information cocycle and ε-bounded codes, Israel J. Math., 29 (1978), 205-20.

[P.7] W. Parry, Finitary isomorphisms with finite expected code-lengths, Bull. L.M.S., 11 (1979), 170-6.

[P.8] W. Parry, The classification of topological Markov chains. Adapted shift equivalence, Israel J. Math., 38 (1981), 335-44.

[P.S.] W. Parry and K. Schmidt, A note on cocycles of unitary representations, Proc. A.M.S., 55 (1976), 185-90.

[P.S'] W. Parry and D. Sullivan, A topological invariant for flows on one-dimensional spaces, Topology, 14 (1975), 297-9.

[P.T.1] W. Parry and S. Tuncel, On the classification of Markov chains by finite equivalence, to appear in Ergod. Th. and Dynam. Sys., 1 (1981).

[P.T.2] W. Parry and S. Tuncel, On the stochastic and topological structure of Markov chains, to appear in Bull. L. M. S., 14 (1982).

[P.W.] W. Parry and P. Walters, Endomorphisms of a Lebesgue space, Bull. A.M.S., 78 (1972), 272-6.

[P.W'] W. Parry and R. F. Williams, Block-coding and a zeta function for finite Markov chains, Proc. L.M.S., 35 (1977), 483-95.

[P'] B. Petit, Deux schemas de Bernoulli d'alphabet denomerable et de même entropie sont finitairement isomorphes, preprint.

[R.1] V. A. Rohlin, On the fundamental ideas of measure theory, Transl. A.M.S. (Series 1) 71 (1952).

[R.2] V. A. Rohlin, Lectures on the entropy of transformations with invariant measure, Russ. Math. Surveys, 22 (5) (1967), 1-52.

[R.3] V. A. Rohlin, Exact endomorphisms of a Lebesgue space, Transl. A.M.S. (Series 2) 39 (1964), 1-36.

[R'] M. Rosenblatt, Stationary processes as shifts of functions of independent random variables, J. Math. Mech., 8 (1959), 665-81.

[R''] D. Ruelle, Thermodynamic Formalism, Encyclopedia of Mathematics and its Applications Vol. 5, Addison-Wesley, Reading, Mass. (1978).

[S] S. Schwartzman, Asymptotic cycles, Ann. of Math., 66 (1957), 270-84.

[S'] E. Seneta, Non-negative Matrices, Allen and Unwin, London (1973).

[S.W.] C. E. Shannon and W. Weaver, The Mathematical Theory of Communication, Univ. of Illinois, Urbana (1963).

[S''] Ja. G. Sinai, Markov partitions and c-diffeomorphisms, Func. Anal. and Appl., 2 (1968), 64-89.

[S'''] F. Spitzer, A variational characterization of finite Markov chains, Ann. of Math. Stats., 43 (1972), 303-7.

[T] S. Tuncel, Conditional pressure and coding, Israel J. Math., 39 (1981).

[V.1] A. M. Versik, Theorem on lacunary isomorphisms of monotonic sequences of partitions, Funk. Anal. i ego Priloz, 2 (1968), 17-21.

[V.2] A. M. Versik, Decreasing sequences of measurable partitions and their applications, Dokl. Akad. Nauk. S.S.S.R., 193 (1970) and Sov. Math. Dokl., 11 (1970), 1007-11.

[W.1] P. Walters, Ergodic Theory - Introductory Lectures, S.L.N. 458, Springer, N.Y. (1975).

[W.2] P. Walters, A variational principle for the pressure of continuous transformations, Amer. J. Math., 97 (1975), 937-71.

[W.3] P. Walters, Ruelle's operator theorem and g-measures, Trans. A.M.S., 214 (1975), 375-87.

[W'] R. F. Williams, Classification of subshifts of finite type, Ann. of Math., 98 (1973), 120-53; Errata, Ann. of Math., 99 (1974), 380-1.

[W''] J. Wolfowitz, Coding Theorems of Information Theory, Ergebnisse der Math., 31, Springer-Verlag, Berlin (1961).

INDEX

Adapted shift equivalence 77
Almost topological conjugacy 69-71
Amalgamation matrix 64
Automorphism 2

Basic identities (entropy, information) 7, 8
Bernoulli
    process, shift 5
    measure 5
Block-code (homomorphism) 11, 54-63
    bounded-to-one 55-59
Block isomorphism 11, 54-63
Bounded codes 35-37
Bowen-Franks invariants 88

Central limiting distributions 36
Coboundary 13, 14
Cocycle-coboundary equation 14, 34
Code
    block 11, 54
    finitary 11, 38
Code-length 50
    future 50
Cohomologous 13, 14
Communication system (of Shannon) 91
Cylinder 3

Division matrix 64

Endomorphism 2
Entropy 7-10
    conditional 7
    of endomorphism 9
    topological 23
Ergodic 2
Exhaustive σ-algebra 3
Extension 2, 3
    natural 3

Factor 2, 3
Filler 42
    entropy 42
    measure 42
    set 42
Finitary code (homomorphism) 11, 38
    isomorphism 38-53
Finite equivalence 64-71
First Čech cohomology group 60-63
Flow equivalence 86-89
Furstenberg's lemma 66

Generator 3
    Markov 72
    strong 3
    topological 72

Herglotz's theorem 16